W9-CLK-351

BLOOM

First Published in 2022 by Cool Springs Press, an imprint of The Quarto Group,
100 Cummings Center, Suite 265-D, Beverly, MA 01915, USA.
T (978) 282-9590 F (978) 283-2742 Quarto.com

26 25 24 23 22 1 2 3 4 5

ISBN: 978-0-7603-7415-3

Digital edition published in 2022
eISBN: 978-0-7603-7417-7

Library of Congress Cataloging-in-Publication Data

Names: Steinkopf, Lisa Eldred, 1966- author.
Title: Bloom : the secrets of growing flowering houseplants year-round / Lisa Eldred Steinkopf
Other titles: Secrets of growing flowering houseplants year-round
Description: Beverly, MA, USA : Cool Springs Press, [2022] | Includes index. | Summary: "Bloom is the only modern guide to cultivating flowering houseplants that includes insider secrets to encourage these beautiful plants to bloom and thrive"—Provided by publisher.
Identifiers: LCCN 2022001195 | ISBN 9780760374153 (hardcover) | ISBN 9780760374177 (ebook)
Subjects: LCSH: House plants—Handbooks, manuals, etc. | Indoor gardening—Handbooks, manuals, etc. | Flowers—Handbooks, manuals, etc.
Classification: LCC SB419 .S7273 2022 | DDC 635.9/65—dc23/eng/20220211
LC record available at https://lccn.loc.gov/2022001195

Design: Megan Jones Design
Cover Image: Lisa Eldred Steinkopf
Page Layout: Megan Jones Design
Photography: Lisa Eldred Steinkopf except by R. Elnore Eldred on page 8 and Chelsea Steinkopf on pages 31 (bottom), 60, 162 (inset), and 170
Illustration: Shutterstock

Printed in China

BLOOM

The secrets of **GROWING FLOWERING HOUSEPLANTS** *year-round*

LISA ELDRED STEINKOPF

The Houseplant Guru

In Praise of BLOOM

"Lisa is the original houseplant influencer. She has paved the way for countless people to feel comfortable having easy-care foliage in their homes. Now, she continues to guide us in adding color indoors with this beautiful book. It inspires, educates, and most of all, encourages a new obsession for plant parents: flowers!"

—Katie Dubow, president of Garden Media Group and QVC guest host

"This is a treasure trove of information that's both up to date and in depth."

—Nancy Szerlag, garden columnist for *The Detroit News*

"As an owner of a houseplant boutique, we get lots of requests for certain plants—low-light plants, beginner-friendly plants, and lately an influx of plants that can be kept indoors that flower. Lisa's book covers everything anyone needs to know about flowering houseplants and is a quintessential guide to anyone looking to add color and perpetual summer to their home."

—Johanna C. Dominguez, owner of Put A Plant On It houseplant boutique in Buffalo, New York

"Lisa's book provides lots of practical tips for growing flowering plants indoors. She provides the detailed plant care information needed for success that is often missing in similar books. Beginner and experienced indoor gardeners will find this book useful. Whether adding new flowering plants to the indoor garden or nurturing existing plants, Lisa's advice will prove beneficial."

—Melinda Myers, author and television and radio host

"The Houseplant Guru has done it again, this time with *Bloom*. Lisa has created an easy-to-read-and-use reference (no matter your plant-parent level of experience) that encourages and guides the reader on how to add a little flower power to your houseplant collection for year-round living color. Those with pets and children will especially appreciate the plant toxicity level rating."

—Maria Zampini, president of UpShoot LLC and co-author of *Garden-Pedia: An A-to-Z Guide to Gardening Terms*

"Houseplants that flower don't get enough celebration because we are so often distracted by their beautiful tropical foliage. Lisa puts blooms in the well-deserved spotlight and walks you through exactly how to enjoy year-round blossoms in your home. In addition to providing all the information needed to care for these plants, I loved the Flower Portraits, which introduced me to so many different blooming houseplants I need to add to my wish list!"

—Maria Failla, host of the *Bloom and Grow Radio* podcast and author of *Growing Joy*

"*Bloom* is a handy resource for new and experienced plant parents looking to add color to their homes with beautiful flowering plants. From tried-and-true varieties to fresh picks that may be familiar to many, this book is sure to help anyone select the right flowering plants for their home (or office) and keep them healthy and beautiful."

—Justin Hancock, horticulturist at Costa Farms

"Lisa is a very well-informed writer and collector who does a wonderful job of sharing her knowledge at a level that everyone can understand. Flowering plants are often overlooked for the home, making this book a valuable addition to any plant lover's bookshelf."

—Jared and Liz Hughes, owners of Groovy Plants Ranch

"I confess! I am far more at home in my outside garden than I am tending plants indoors. From orchids and African violets to amaryllis and kalanchoe, I need help. And now I have it. Lisa always explains what to do in an easy-to-understand style. In this book, she also explains why I should be doing it. The in-depth discussions are enlightening, and the plant-by-plant guides are invaluable."

—Allan Armitage, award-winning educator, horticulturist, and author of *Armitage's Garden Perennials* and many other books

This book is dedicated to the memories of two amazing women: my mom, Christine Ann (Baldwin) Eldred, and my grandma, R. Elnore (Lyon) Eldred. These two were wonderful influences in my life and instilled in me the love for houseplants. I miss them.

CONTENTS

ORANGE *GUZMANIA* (PAGE 59)

PURPLE AFRICAN VIOLET (PAGE 105)

PHALAENOPSIS (PAGE 123)

AMARYLLIS (PAGE 131)

Introduction

Are the only flowers you've ever had in your home the type that come in a plastic sleeve, accompanied by a box of chocolates and a card? That's a wonderful way to have flowers for a week or two in your home, but don't you wish you could have flowers in the house all year long, without buying bouquets of them? Or maybe you have purchased flowering houseplants, only to never see a blossom again after the initial bloom. Or maybe the buds all fell off as soon as you got them home. This can be discouraging. Yet, there are plants out there that will bloom in your home without too much fuss. Foliage plants certainly have their place as the backbone of any houseplant display because they are the perfect foil for flowering houseplants. Green plants, pink plants, variegated plants, and black plants are all lovely, but if you are looking for something a bit more exciting and colorful, then it's time to bring flowers into your life all year long with flowering houseplants.

My mom always had her fern in the east window with an enormous spider plant hanging next to it. I still have that fern today, now over sixty-four years old. And I remember when I was younger, riding my bicycle to Grandma's house a half mile (1 km) away and seeing her east windowsills filled with African violets in bloom along with her foliage plants. I love my fern, but those African violets were so beautiful with their pink and blue flowers—and it seemed they were always blooming. I wish Grandma could see the variety of African violets there are today because she would be amazed. She fussed over those plants, making sure they had everything they needed, and she always had a leaf or two rooting in an aluminum-foil-covered baby jar. I was young and didn't pay too much attention to the propagation process, but there were

Grandma Eldred's African violets were always in bloom on her east kitchen windowsill.

always violets blooming; those small flowers helped me learn to love flowering houseplants and gave me the desire to always have blooming plants on my windowsills.

One of the most popular flowering plants (or maybe I should say most well-known?) is the peace lily, or *Spathiphyllum*, but there are so many more plants that will bloom in your home. Peace lilies are nice, and African violets are one of my favorites, but don't limit yourself to just a couple of varieties. You are much more likely to have blooms all year long if you have several types of plants.

In this book, I want to show you plants I have had success with or have witnessed others' success with. One plant may do better for you than for someone else because you have better conditions for that plant. Providing your plant with the environment it needs to thrive makes all the difference. If you want a cactus to bloom, you should have an unimpeded south or west window or have it placed under grow lights. If you don't have a southern window but do have an east window, you still will have many plants to choose from that don't require full sun to bloom. I want everyone to be successful growing flowering houseplants, and hopefully, to get one or more to bloom regularly. Let's dive into the world of flowering houseplants and learn about ways to meet their needs so you can have flowers all year long.

Caring for Flowering Houseplants

Care for flowering houseplants is a bit different than care for their nonflowering counterparts, the main difference being that flowering houseplants need more light to bloom than plants grown for their foliage. Blooming takes a large expenditure of energy from a plant, and that is one reason plants that are grown exclusively for their foliage are often disbudded if and when the plant begins to bloom. Most plants develop flowers in nature, and we may never be aware of that as they are sold exclusively as foliage plants and usually don't receive the conditions needed to bloom in our homes. So, their flowering ability isn't considered when they're grown for the houseplant consumer.

A good example of this is the *aglaonema*, or Chinese evergreen. It blooms like other plants in the aroid family, but I cut the flowers off because I don't want any energy taken away from the production of their beautiful leaves. Plants send out flowers so that through the production of seeds, they can replicate themselves. Often the seeds are surrounded by a fruit that can be consumed, such as an orange or apple. These fruits are there to entice animals to eat them and spread the seeds further afield for the plant. If you aren't interested in starting seeds or if you have a desire to eat the fruit or berries, cut the flowers off. Yet it is exciting to have a foliage plant bloom when it normally doesn't bloom, and it's okay to leave the flowers on if you so choose.

There are a few things that need to be addressed when caring for flowering plants. It is one thing to get them to bloom, but it is another thing to keep them flowering for as long as possible. In this chapter, we will talk about ways to care for your houseplants before they bloom, care for them while they bloom, and care for them after the blossoms have dropped. There are many factors to consider, including light; the container and potting medium; fertilizer; humidity; and most importantly, the correct watering practices. A well-tended plant is more likely to be healthy, and a healthy plant is more likely to bloom.

The origins of a plant are clues for its care

When the native habitat of a plant is mentioned in this book and elsewhere, there is a reason for that. Though we might not have been to that region, by looking up the climate and growing season of that area, we can find clues to the care and growing habits of our houseplants. We obviously can't bring the rainforest climate to our entire house, but we can emulate it to a degree by adding humidity or even growing that particular plant in a terrarium or enclosure that is more conducive to their health. Many people collect specific plant families and are willing to go to great lengths to create a favorable environment to ensure the health of their plants.

LIGHT

Flowering plants are most often purchased when they are in bloom. When the flowers are spent, they are cut off, and the plant is then diligently cared for in hopes that it will send out flowers again for future enjoyment. Unfortunately, sometimes this doesn't happen and questions arise about why flowering plants never bloom again after their initial bloom. In most cases, it is because the plants aren't receiving enough light to trigger them to bloom. There are other factors to consider, but light is the first condition to address. Light is the life-giving food that plants need. Try to determine if the nonblooming plant is getting the correct amount of light to allow it to manufacture enough energy to produce flowers, or if it is only receiving enough light to sustain its leaves.

An *anthurium* is basking in the sun on this windowsill.

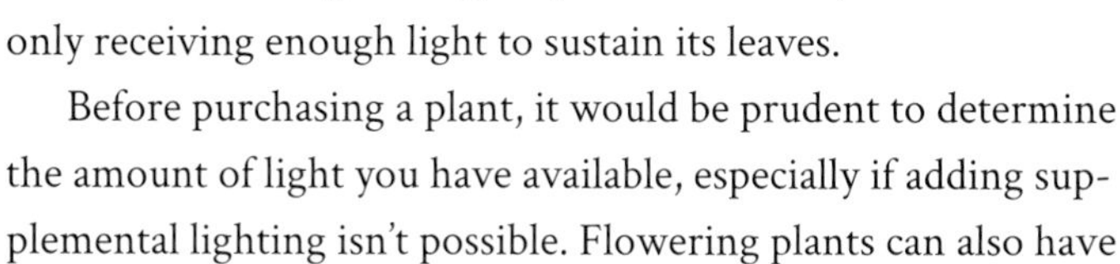

Before purchasing a plant, it would be prudent to determine the amount of light you have available, especially if adding supplemental lighting isn't possible. Flowering plants can also have beautiful foliage, so if there isn't enough light to trigger flowering and supplemental light is not available, you can always enjoy your plant for its foliage. If you only purchased the plant because of its blooms, gift it to a friend with stronger light conditions or consider investing in a grow light (more on these later).

Ambient Light

Let's discuss the ambient lighting already available in your home. First, it is necessary to decide which exposure your windows face: north, south, east, or west. Remember to check outside for objects that will block the light, such as trees, shrubs, or the neighbor's home. Most flowering plants are going to require medium to high light conditions to flower. As a result, many plants will not flower in a true northern exposure. Begonias, jewel orchids, and gesneriads (members of the African violet family) love the soft morning light of an east window and will consistently send out flowers if their other needs are met. A south or west window is necessary for most desert cacti, but jungle cacti, such as Thanksgiving or Easter cacti, will thrive in an east or west window. If you live in the southern hemisphere, these directions are reversed. What is the right amount of light, you might ask? As you can see, it depends on the type of plant you want to grow and what environment is best for them to stimulate flowering. That is the first thing to determine.

All plants have yearly cycles of growth and rest, whether it be from a dry period or lower light levels. Within that yearly cycle, a flowering plant should bloom. If it doesn't perform within a year's time frame, try to decide what is hindering its flowering capabilities and rectify the problem.

The light is perfect for this *Phalaenopsis* orchid to bloom and thrive.

Does your nonblooming plant need to be moved to a different window with more light? Even if it is in the sunniest window, it may need a grow light to boost the light levels even more. Not everyone has windows that meet the needs of the plants they want to grow, so supplemental light must be added. You can grow and flower any plant under electric lights. With the correct lights, a windowless room can produce beautiful plants and flowers. There are currently many types of lights to choose from, and finding the right one for the plants you want to grow is important. There are lights available for desktop plants as well as multilight shelving systems for collections of shorter plants and lights that can be suspended over larger plants. With so many choices, it may seem a daunting task to choose a grow light that is right for your situation. The key is to know the type of plant and the number of plants you want to grow.

Grow Lights

There are numerous grow lights to choose from now, but in the not-so-long-ago past, the choices of lights for growing plants were limited. Incandescent bulbs, like the ones used in regular household lamps, and T12 fluorescent bulbs were the most common ones used in the home. Now, along with those bulbs, there are LED (light-emitting diode) lights, which are far more energy efficient and cost-effective in the long run. Which lights you choose to provide or supplement the ambient light is up to you and your plant's performance under them. Many growers find that fluorescent lights are still a great cost-effective option. The compact fluorescents are a good fit if you have just one plant that you need to illuminate, as the light from one bulb won't spread far enough to support more than one plant. Replacing all your incandescent bulbs in your lamps with compact fluorescent bulbs will be a bit helpful to nearby plants, depending on how far the plants are from the light. More importantly, the fluorescent bulbs have a wider range of light color options, are more energy efficient, and give off less heat than an incandescent bulb—which, if you've ever changed a light bulb, you know can be finger-burning hot.

There are three sizes of fluorescent lights. They come in many lengths, but it is the diameter size that indicates which light you are buying and the fixture it will need. The T12 bulb was the one most commonly used for "office" light fixtures, and it works fine for plants, but now, the more efficient T8 and T5 bulbs are taking its place. The T12 bulbs are the largest fluorescent bulbs at 1.5 inches (4 cm) in diameter, the T8 bulb is 1 inch (2.5 cm) in diameter, and the T5 bulb is ⅝ inches (2 cm) in diameter, so clearly they have become smaller and at the same time more efficient. These smaller fluorescent bulbs work well for plants and are still widely used today.

Using electric lights will give your plants the same amount and duration of light every day so they will bloom more reliably.

The newer kid on the lighting block is the LED bulb, and though they are coming down in cost, they are still the more expensive choice. In the past, these bulbs were exclusively available in red and blue types, casting your plants in a pinkish-purple light. Those two colors may seem like a weird combination, but light wavelengths in the blue and red ranges are the colors plants require to grow, flower, and thrive. Blue light helps with foliage growth, and red light is helpful for flowering. Plants growing under only one of the red or blue colors will suffer. Combined red and blue color spectrum lights are great for plants and are highly energy efficient, but obviously they will distort the color of the plants to the human eye.

Luckily, now there are LED lights that provide the full spectrum of light and won't change the color appearance of your plants to any great extent. Long tubular LED lights or square panels of LED lights work well on stands, illuminating a large collection of plants. If you have one large plant that needs supplemental light and is standing on its own in your home, don't despair; there are hanging spotlights that will supply it with plenty of light to thrive.

HOW LONG TO LEAVE THE LIGHTS ON

One of the most important factors of having a "light garden" is the amount of time your plants are illuminated. Plants will need between seven to ten hours of LED light exposure or ten to fourteen hours of fluorescent light exposure. Don't forget there are short-day and long-day plants that require different lengths of light and darkness to bloom as well as day-neutral plants that like an equal amount of light and darkness (more on this in a later section).

HOW CLOSE TO PLACE THE LIGHTS

Placing a plant the optimum distance from the lights is also vital. It may grow best 6 inches (15 cm) under the lights or 24 inches (61 cm) away, depending on the type of plant, its size, and the light levels it prefers. A cactus or other succulent-type plant will need to be closer to the lights than an African violet, for example, yet they could both be grown on the same stand at differing heights below the light fixture.

This spotlight is a powerful grow light that can illuminate one large floor plant, such as this gardenia standard, or a few smaller plants.

This 'Disco Dancer' African violet is being grown under LED lights.

OTHER FACTORS TO CONSIDER

There are visual indicators that will help you determine if your plant is doing well in the light it is exposed to. If your plant is stretching upward and reaching toward the light, it needs to be moved closer. If plants have tightly bunched or sun scorched leaves, are excessively compact, hug the pot, and draw away from the light, they may need to be given more distance from the light.

What Can Go Wrong

Phototropism and sunburn are light problems you may encounter. If you have a plant that hasn't been turned and is leaning toward the window, this response is called phototropism, which is defined as movement toward light. There is positive tropism, which means it turns toward the light, and negative tropism, which means it moves away from the light. Yes, plants can move away from light if it is too bright. A good rule of thumb is to turn your plants one-quarter turn every time you water so all sides get equal light exposure and the plant grows symmetrically. When it pertains to flowering, this is especially important because if only one side of the plant is exposed to the light, there may only be blooms on that side. Note that even plants growing under fluorescent light should be rotated because light intensity varies throughout a lightbulb.

Sunburn on plants is something most people don't think about. All plants have a cuticle or waxy covering on their leaves that helps to keep moisture in the leaf and protects it to some extent from disease and insects. It can be compared to using sunscreen on our skin. In our homes, a thick cuticle layer on plant leaves isn't necessary as the light levels are much lower than outside. In the summer, plants are often moved outside, and sunburn can even affect cacti, other succulents, and some high light plants. Though they thrive in full sun in their native habitats, high light plants have become accustomed to living in our lower light homes. When any plant is taken from inside to outside, moving it to a higher light situation

A timer, whether the old-fashioned kind or a plug that pairs with an app on your phone, will turn the lights on and off for you.

Using a timer

To increase the efficiency of an electric lighting system, add a timer so the responsibility of turning the lights on and off every day is not yours. Even if you are away from home, the lights will turn on and off automatically, giving the plants the same amount of light every day. You may choose to change the timer settings throughout the year, especially if you are growing a short-day or long-day plant (more on these plant types in a later section). Remember, with any houseplant, all the care is up to you.

should be done gradually. This can be done by putting the plants under a shady tree or on the north side of the house without any direct sun. Leave them there for a couple of weeks so they can build up their leaf cuticle to better withstand the sun. This is called acclimating your plants. And before they are brought back inside in the fall, place them in the shade again so they can slowly get used to living in a lower light situation again. Doing this keeps your plant from getting damaging sunburn and reduces the shock.

This Thanksgiving cactus wasn't turned throughout the year, causing it to bloom heavily on only one side.

Plant hunters

If it weren't for plant hunters in previous centuries, we wouldn't have many of the plants we have today. These bold, courageous explorers braved disease, sickness, wild animals, and voracious insects to hunt for the plants that today we call houseplants. They sailed to strange lands with indigenous people who often weren't amenable to strangers invading their land (and rightly so). Many plant hunters named their plants after people or the areas in which they were found. I have tried to find either the explorer who found the plant or for whom the plant was named and included this information in the plant profiles found in chapter 3.

This leaf was damaged when the plant was moved outside into the direct sunlight without being acclimated.

POTTING MEDIUM

Though light is extremely important for the flowering process, it isn't the only factor that contributes to a plant's blooming. A significant component to a robust flowering plant is the potting medium in which it is growing. Healthy plants need healthy roots, and the medium the plant is growing in makes a difference.

There are countless potting mediums to choose from. Serious collectors concoct their own growing mixes, and some even keep the ingredients, or at least the proportions, under wraps. It isn't necessary to take matters to that extreme because there are commercial mixes readily available that are formulated specifically for certain plant families. There are many plant parents who purchase these specialty mixes and then add amendments, tweaking them to their liking and according to what they perceive to be best for their plants. Even in the cacti family, for example, there are different genera and species that live in different environments that would require more specific components. There are African violet mixes that will work for other gesneriad family members and growing mixes for citrus and palms as well.

Choose the potting medium that is best suited to the plant you are growing.

If you choose to add ingredients to your purchased mixes, there are numerous choices to enhance the mix for your specific plant. I suggest that if you decide to add ingredients to make your own mix, wear a dust mask for safety and perhaps gloves.

Potting Soil Amendments

Some of the most common and readily available amendments are vermiculite and perlite. Less common options include pumice, grit, Akadama, and lava rock. Let's discuss these amendments and how they can help improve your potting medium.

Vermiculite: Vermiculite is a mineral that is mined from the ground. After it is heated, it expands and becomes a flaky, shiny material. It has been used in the horticulture industry since the 1960s to improve aeration in the soil, and it holds on to nutrients and releases them to the plant as needed.

Perlite: Perlite is a volcanic glass that, like vermiculite, is mined from the ground and is heated so that it expands in size. It improves the drainage, doesn't hold a lot of water, and helps prevent the medium from becoming compacted over time. One drawback to using perlite is that it floats and turns brown over time. Many find it unattractive and dislike the floating aspect, so some substitute it with comparable amendments such as pumice or grit.

Pumice: Pumice is a mined product like vermiculite and perlite but is less readily available. It is a volcanic glass rock that spews from a volcano when it erupts and forms holes as it rapidly cools. It is lightweight like perlite, assists with drainage, doesn't compact, and doesn't float. Its many pores allow for air exchange and nutrient storage.

Starting from the left, a purchased potting medium is mixed with vermiculite and perlite to make it a more well-drained mix.

Grit: Grit is a product used for chickens and turkeys that can be found at a local farm supply store or online. Though it adds weight, it helps with drainage and is often used in cacti and other succulent mixes. Top-heavy plants, such as large cacti and many succulents, are made more stable with grit in the mix.

Akadama: Akadama is a Japanese red claylike material most often used in a bonsai potting medium. Akadama is heavy and hard to procure locally. If buying online, shipping costs will be higher because of the weight. It provides aeration and aids with drainage while having good water retention at the same time. Many people are beginning to use it exclusively as their growing medium.

Lava rock: Lava rock is created by volcanic eruptions and is crushed into small pieces. Many remember this being used as outdoor mulch in the 1970s and 1980s; some may have even bought a home and had to remove it. Removing it was not easy, but as a soil amendment, it works well aiding with drainage.

Amendments that could be added to a potting medium include: Back row (left to right) Akadama, perlite, vermiculite, and a succulent mix containing coir, lava rock, grit, and pumice. Front row (left to right) worm castings and orchid bark.

Coir: (pronounced COY-er) Coir is used as a substitute for peat in potting mixes. It is made of fiber from the husks of coconuts and is used in products like rope and doormats. Because peat is in question as a nonrenewable resource, coir is a good replacement as it is a byproduct and comes from a food plant that is being grown all the time. It holds water like peat and is readily available.

Orchid bark: Orchid bark is usually bark from fir trees and can also be used as an amendment to a potting medium. It will assist with drainage (yet still holds some water). There are different sizes of orchid bark, and if it is used as an amendment in your medium, use a smaller sized bark.

Worm castings: Worm castings are just what you think they are—worm poop. This product is also called vermicast, vermicompost, or worm compost. Worm castings are full of minerals that are slowly released to your plant when added to the growing medium. Mix worm castings into the potting medium in small amounts when repotting your plants.

Epiphytic plants vs. parasitic plants

The word *epiphytic* will come up in a few of the plant profiles in chapter 3, and it may be a confusing concept, so I'll explain it here first. Tillandsias, jungle cacti, and many orchids grow epiphytically in their native habitat. But what is an epiphyte?

An epiphyte is a plant that makes its home on another plant rather than in soil, yet it isn't taking any nutrition from the plant it's growing on. Epiphytes typically have a modicum amount of area in which to root, usually in the fork of tree branches. That is the reason most epiphytic plants prefer a potting medium that is coarse, where water runs through it quickly. These plants usually receive their nutrition from animal droppings and decomposing leaf debris from the plants they are living upon.

A parasitic plant also grows on another plant, but unlike the epiphyte, it draws energy from the plant it has chosen to infringe upon. Eventually, it may lead to the host plant's death. A common example of a plant that is living on and stealing energy from another plant is the ubiquitous holiday plant that entices couples to kiss, the mistletoe.

CONTAINERS

There are countless styles, materials, and sizes of containers to use for your plants. The most important factor is that the container has a drainage hole. This is especially vital if you are a new plant parent. Though there are many examples of plants living their best lives without a drainage hole, this is an anomaly. Knowing what is going on in the bottom of the pot is important. With a drainage hole, it is obvious that water is getting to the bottom of the pot because it runs out the hole. Whether or not the entire root ball is getting hydrated in the process is another factor, but we will talk about that more in the watering section. If your pot doesn't have a drainage hole, it is an easy thing to remedy, using a diamond-tipped or masonry drill bit.

These two types of bits will allow you to drill a hole in almost any material so you can have drainage. If you prefer not to drill a hole, use the container as a *cachepot* (French for hide-a-pot). Keep the plant in its utilitarian grower's pot and "hide" it inside the cachepot without a drainage hole. Take the plant out to water, allow it to drain, and return it to the cachepot.

If a plant is in a cachepot like this orchid, remove it, water at the sink, drain, and return it to the cachepot.

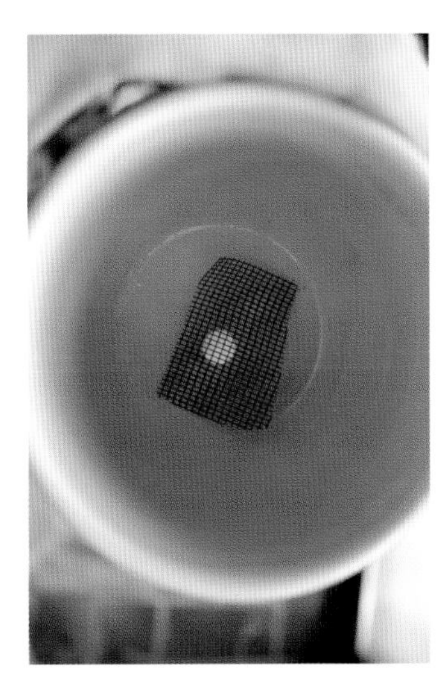

If your container doesn't have a drainage hole, it is easy to drill one with a diamond-tipped or masonry drill bit.

Pot Materials

The kind of pot used is also a factor to consider. If you have a plant that prefers to be on the dry side, such as a cactus or other succulent, using a porous clay pot is recommended. That isn't to say that they won't do perfectly well in a glazed container. Remember, you are in control of the amount of water your plant receives. The porous wall of the clay pot allows water to escape not only through the drainage hole, but through the wall of the pot as well. If, on the other hand, a plant prefers to stay moist, a glazed container would be a better choice.

There are so many choices of containers. First, let me say, choose a container for your plant that not only suits your plant but also makes you happy. If you only like brightly colored pots, pots that match your décor, or white pots, that's fine. Use them.

Pot Shape

The shape of the pots should also be considered. Many plants have shallow root systems, including succulents and various gesneriads, and prefer a shallow pot, such as an azalea pot (the depth is three-quarters of the diameter) or a bulb pan (the depth is one-half of the diameter). Some of these containers can be challenging to find, but they are worth the effort for special plants. The container shape that should be avoided is the bulbous one where the middle part of the container is wider than the top opening. Though they are attractive, getting a plant out of them to transplant it is almost impossible without cutting away a large portion of the root ball.

There are many types of containers and saucers to choose from.

Shallow pots are perfect for plants with shallow root systems.

WATER

It is often reported that water is the biggest killer of plants. That may be true, but it's the person watering it who is actually the culprit. Either water isn't applied often enough, or it is applied too often. Many new plant parents have watering questions such as, "When do I water? Should I water all my plants once a week? How much do I give each plant? One cup, two cups, or just a quick tip of the watering can?" These are frequently asked questions, especially for new plant parents.

Watering on a schedule can cause problems, yet it works perfectly fine for some people. A strict watering schedule doesn't take into consideration the amount of sun received by your plant between waterings. A cloudy week would prohibit your plant from photosynthesizing as much, thus it would not use as much water. Maybe it was an exceptionally hot, dry week and the air conditioner was working overtime, cooling your home, and removing all the humidity. A plant would use more water under those conditions, so it may need extra water before the next scheduled watering time. It is better to **check** your plant on a schedule, but not necessarily to water according to a scheduled set time.

Using cold water can damage sensitive leaves and may shock the roots, so use tepid or room temperature water for your plants. Damage on an African violet leaf is shown above.

Bottom watering allows the water to be drawn up into the entire root ball.

How to Water Properly

Overwatering a plant simply means that water is applied too often, not allowing the plant to dry out between waterings. Unless it is a true aquatic plant, there aren't many plants that need to stand in water.

It is necessary to water a plant thoroughly every time water is applied, which means watering a plant until water runs out the drainage hole. It is challenging to tell if water has completely saturated the soil all the way to the bottom. For larger pots, insert a dowel to check the moisture level at the bottom of the pot. For smaller pots, lift the watered plant or use your finger. The heavier the pot and soil, the wetter it is. This is important because the whole root ball needs to be moistened.

Bromeliads that grow in a vase shape need water to be added regularly to their tank (cup) to keep it filled.

When checking a large pot, insert a dowel to check if the plant is moist at the bottom of the pot.

If only enough water is added to wet the top half of the soil, the lower roots may be damaged from desiccation. This is also true if a plant is only ever watered in the same area of the pot. When applying water, it should be evenly distributed around the pot so that all the roots are moistened. If pockets of the root ball are left dry, those roots could die and consequently you may lose the part of your plant that those roots supplied with water. This is common when a larger plant is involved. If your root ball has dried out excessively and the potting medium has shrunk away from the sides of the pot, water may just run between the medium and the pot, never wetting the medium at all. If this occurs, bottom watering would be more conducive to rewet the entire root ball. When the water runs through the drainage hole and into the saucer below the plant, let the plant take up the water for an hour, then empty any excess water from the saucer. If your plant is too large to move, a turkey baster works well to remove any leftover water.

Use a turkey baster to remove excess water from the saucer of a plant that is too large to move.

Aerial roots

We all know about the roots that are in soil and support our plants by anchoring them and providing nutrients, but what are aerial roots? Aerial roots are roots that are above ground. They are usually part of an epiphytic plant's makeup, though some vines such as monstera and English ivy also have them. The most common aerial roots are the ones that are seen on orchids. The roots help anchor these plants to whatever host they are living on and also collect water from humidity, dew, or rain. Aerial roots (particularly on many orchids) may have a green covering called velamen that makes them able to absorb water from mist in the air. And because the velamen contain chlorophyll, the aerial roots can also photosynthesize for the plant.

HUMIDITY

In their natural habitat, many of our houseplants are accustomed to elevated levels of humidity. Many are found growing in rain forests that have extreme humidity and balmy temperatures most days of the year. There are no true "house" plants. When we bring plants into our homes, which are centrally heated and/or cooled, most are being brought into new environments they are not going to like. However, plants are adaptable and will for the most part acclimate to our homes. It is best if we recreate their natural habitat to the best of our ability with the resources we have access to. If a plant needs high humidity, consider growing it in a terrarium, add a humidifier to the room, or, at the very least, place your plant on a pebble tray.

Humidity is especially important for flowering plants because low humidity can be a factor in why a plant isn't blooming. Consequences may include buds falling off before opening, flowers that don't last as long, and the edges of the petals turning brown. Though misting plants may be therapeutic to you, it doesn't accomplish much for our plants in the long run. The mist created will quickly dry, and the plants will be back to their original humidity levels.

Set your flowering plant on a pebble tray to elevate the humidity around it, especially while it is blooming.

Though misting may be therapeutic for the plant parent, it doesn't raise the humidity level for a length of time beneficial to the plant.

GROOMING

Use a soft brush to dust fuzzy leaves or spiny plants that are difficult to clean.

Grooming is something often overlooked by plant parents because they assume plants are pretty much self-sufficient in this arena. They grow leaves, lose leaves, make flowers, drop old flowers, make seeds, and more, all by themselves. That is true, but fallen leaves would naturally decompose in their habitat, so in our homes, discarding plant debris before a fungus moves in or insects find a new hiding place is a good idea. It also doesn't rain in our houses, so manually cleaning plant leaves is important. Dusty plants aren't photosynthesizing to their fullest ability. A clean leaf means that all the light can reach it, allowing for more efficient food production. Plants make their food from the energy of the sun (or electric lighting), and they also need water and carbon dioxide. Keeping the leaves clean helps this process.

Take your plants to the sink or shower and give them a gentle rinse to remove all the grit and grime from their leaves. If there are plants growing in or near where cooking is being done, they may need to be cleaned more often since a greasy buildup on the leaves can become a problem. Cleaning windows and screens often can also make a difference in the amount of light your plant receives.

Cleaning cacti and other succulents that have spines can be tricky. They are fine with a shower cleaning like other houseplants, but if you have a cat or dog, the fur often stubbornly clings to the spines. A small, soft paintbrush or makeup brush works well to lightly brush off dust and fur. This also works well between waterings or in the winter when completely saturating your cacti roots may be detrimental.

leaf-cleaning tip

Use a cut lemon slice to get hard water spots as well as any greasy residue off the leaves. Full-strength lemon juice is fine for most plants, but you may want to dilute it a bit with water. Wipe the leaves down and rinse them well; your plants will thank you.

If your leaves have mineral spots on them, cut a lemon and wipe the leaves with it. Rinse and the leaves will be all clean!

Keeping Pots Clean

Having clean and dust-free plants is important, but don't forget to also clean their homes. Residue from fertilizers, dust, minerals in the water, and the potting medium can build up on the pot, forming a scaly substance. This isn't easy to get off, but if your pots are wiped down often, this should help prevent that problem. Even if your plant isn't ready for a new pot, it can still be removed and set aside while you are cleaning the container. Don't forget to clean the saucer as well, which often becomes extremely dirty and could be harboring insects and disease. Vinegar works well to remove that stubborn scaly residue, but always wash and rinse the container well before placing a plant back in it.

It may be necessary to stake a flower stalk. Use a dowel and small clip or soft tie to secure the stalk and prevent damage.

Keep dead leaves and roots removed, as on this orchid, so that disease and fungi do not spread.

Pruning

Often our plants outgrow the area we have to offer them, so pruning is needed. Not only can you prune the branches of the plant, but you can utilize root pruning in order to keep the plant smaller without needing to switch pots. If a plant is root pruned at the same time the top is pruned, the plant can stay in its existing container even if you were thinking of moving it to a larger home. It's best to prune the top at the same time as making the root system smaller because if you don't, the roots will be unable to support the entire plant above it. As you prune your plants to keep them in check, keep those cuttings so you can start new plants to share or expand your collection.

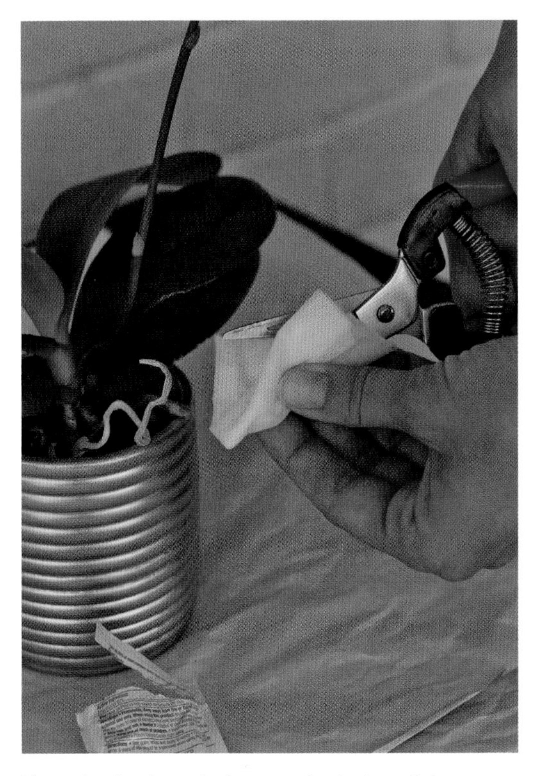

Keep tools clean between plants by wiping them with isopropyl alcohol.

Cuttings can be rooted in water, like this small, hanging test tube propagator.

It's Bloom Time!

Do you have a flowering plant that refuses to bloom? Have you tried all the conventional and maybe some not-so-conventional ways to try to coax it into flower—fertilizing heavily, repotting to a larger pot, root pruning, cold temperatures, drought, and even begging it to bloom—all to no avail? Most flowering plants should flower within the span of a year's time. All of nature is cyclical, so if you feel you are meeting all the demands your plants need to send out flowers yet they aren't, there's still some requirement that is not being met in its current environment. Let's address some conditions that affect your plant's blooming processes and see if we can help your nonblooming flowering plant burst into bloom.

BLOOM CYCLES AND PHOTOPERIODISM

I'm going to get a little technical for a moment here, but bear with me. It's important information, I promise. Most flowers bloom in a certain period each year, whether it be spring, summer, fall, or winter. Of course, if your plants are growing under electric lights, this may change their natural cycle that would normally follow the outdoor seasons. Plants that bloom in the late fall and winter when daylight is fleeting and the nights lengthen are referred to as **short-day plants.**

Those plants that bloom in the spring and summer when the days are lengthy and nights are short are specified as **long-day plants**. The plants that are not concerned with the span of light and dark are **day length indifferent** or **day-neutral**. These neutral day length plants grow near the equator, where the amounts of light and darkness are the same every day.

Poinsettias and holiday cacti are short-day plants.

Use ripening fruit to promote flowering

Ethylene gas is a gaseous plant hormone used to promote flowering in stubborn non-flower producing plants. It is often used in the commercial houseplant industry to help get plants to flower when they are not doing so on their own. This method won't work unless the plant is mature enough to flower.

In your home, this encouragement can be accomplished by using ripening fruit such as an apple, which releases the required ethylene gas. Place your plant in a clear plastic bag with the apple for a week, and if all other factors are met to promote flowering, your plant should flower within a month or two.

This phenomenon caused by a response to the length of light and darkness is called **photoperiodism**. The word comes from Greek roots meaning "light" and "length of time," but in reality, it is the amount of darkness that is important (see the sidebar on page 41 to learn more about which plants are influenced by photoperiodism). Another key to this phenomenon is termed **critical day length**, or the measure of day or night hours necessary in a twenty-four-hour period to prompt flowering of long-day plants or to keep short-day plants from flowering at the wrong time. Long-day plants require light for a longer time than their critical day length, and short-day plants require light for a shorter period than that critical day length. The phenomenon of plants responding to light or photoperiodism was first discovered and written about by W.W. Garner and H.A. Allard in the early 1920s.

African violets are day-neutral plants.

One may assume that plants might get spring and fall mixed up. The light is getting longer in the spring and shorter in the fall, so that might trigger flowering, and sometimes it does. Yet most plants do not get these two seasons mixed up because they are either getting warmer in the spring or cooler in the fall. Some plants not only react to light differences but to temperatures as well. If plants need a cold period before blooming, this is called **vernalization**. Many cacti bloom in the late fall or early spring after experiencing colder temperatures and shortening or lengthening days. Short-day cacti, such as the Thanksgiving cactus, react to the shorter days of fall along with the cooler temperatures that occur at that time. If your plant is experiencing cooler temperatures and shortening days, naturally, water is applied less often. This period of dryness may also help to trigger flowering.

In commercial growing settings, plants are treated with chemical growth regulators and/or light manipulation to achieve flowering when the grower and the public want/need the plant to be in bloom. The same plants may not bloom at exactly the same time in your home the following year because you won't be using chemical growth regulators or strategic light manipulation to trick the plant into flowering when you want it to. These tactics are used to ensure holiday plants bloom when the consumer is most likely to be purchasing them.

Photoperiodism—which plants use it?

Photoperiodism is a plant's response to changes in day length. Different plants respond differently to day length, and they can be grouped into three types. These include short-day, long-day, and day-neutral plants.

For our purposes, we will talk about how photoperiodism pertains to the flowering capabilities of plants. The amount of uninterrupted darkness is the key to flowering. A plant whose flowering is triggered when nights are long is called a short-day plant, and this group includes poinsettias, holiday cacti, rhizomatous begonias, and kalanchoe. Greenhouse growers black out their greenhouses for fourteen to sixteen hours of complete darkness every night for many weeks to trigger these plants to flower. If the darkness is interrupted at night with light, the affected plants won't turn their beautiful poinsettia colors and will stay green. Some home gardeners will place poinsettias or holiday cacti into a dark closet for fourteen to sixteen hours each day for ten to twelve weeks to encourage them to flower in their homes.

A plant whose flowering is triggered when nights are short are called long-day plants, and these include most of the flowering annuals we grow outside.

Lastly, there are plants that don't care about the amount of darkness they receive, and those are called day-neutral plants. These include African violets, streptocarpus, and most gesneriads. Plants that grow near the equator where days and nights are the same are usually day-neutral.

FERTILIZER

As I mentioned in the lighting section, the only thing that can enable your plant to bloom is the light it receives, whether from the sun or electric lights. Each plant has a quantity of light it needs to stimulate the plant to bloom and will need different amounts of light depending on where they grow in their native habitat. There are plants growing as epiphytes in the treetops, receiving dappled sunlight, whereas other plants grow in the full blazing sun and may never bloom in our homes, as we cannot provide the levels of light they need. Fertilizers are often labeled as plant food. Though that is the perception companies want the consumer to have, fertilizer is not food for your plant; light is. I compare fertilizing your plants to humans taking vitamins. It isn't necessary, but it does help provide nutrients and minerals we may not get from our food intake.

Fertilize your plants following the label instructions; you may use less than called for, but never more.

Where do plants in nature get their nutrients? Epiphytic plants, such as orchids and bromeliads, get their nutrients from the debris that falls and lands in their root area, or in the case of vase-shaped bromeliads, into the cups where they collect their water. Animal droppings are also a good source of nutrients. Terrestrial plants receive nutrients from the soil they are growing in. Leaves drop, branches fall, and the decomposition of those natural items provides nutrients to the plant. There are also earthworms and other animals that stir up the earth and leave "fertilizer" behind for the plants.

If a plant hasn't bloomed, it is often assumed that fertilizing the plant will make it bloom. Though fertilizer isn't the food source supplying sustenance to your plant, it doesn't mean that it won't help your plant bloom. Adding fertilizer to your plant care regime is good for your plant. A healthy plant is more likely to thrive and, in the case of flowering plants, bloom.

Reading the Label

There are fertilizers specifically labeled for flowering plants. These fertilizers have a higher middle number than the other two numbers on the label. The three numbers stand for the three macronutrients that are most needed by a plant and represent the percentage of that nutrient contained in the fertilizer. There is an old saying that can be helpful when trying to remember what the three numbers stand for: "up, down, all around."

- The first number stands for the percentage of nitrogen (N) in the fertilizer. Nitrogen helps with the green or "up" part of the plant. It promotes healthy leaves and new growth.
- The middle number is the percentage of phosphorus (P). Phosphorus helps with the health of the roots or "down" part of the plant and the blossoms.
- The last number is the percentage of potassium (K). Potassium helps with the "all around" health of the plant.

Though some fertilizers especially formulated for blooming plants imply they will make a plant bloom, they will not. It will, however, help the flowers grow larger, cause blossoms to be more vibrantly colored, and help the pedicels grow stronger to hold up the flowers. It is a booster for the flowers, not a maker of flowers.

Types of Fertilizers

There are numerous types of fertilizer to choose from, and you should use the one you find works best for your plants. Whether it be slow release, water soluble, or an organic fertilizer, any is better than none. Always follow the directions on the label, never using more than is listed on the label for safe use. Applying too much fertilizer could burn your plants and even injure them to the point of no return. I never use the full strength called for on the label; instead, I use three-quarters to one-half of the recommended amount. If you prefer, you could use one-quarter strength every time you water your plants during their active growing season.

Add water-soluble fertilizer to your watering can per recommendations on the label.

How Often to Fertilize

This brings us to the question of how often and within what time frame fertilizer should be used. The rule of thumb has always been once a month, whereas I like to suggest every fourth watering. That may be more or less often than once a month, depending on how often the plant requires water. The time of year is also key, and fertilization should only occur when plants are actively growing. If your plants are residing under electric lights or you live in a tropical climate, fertilize year-round. If they are grown in natural light, it is best to withhold fertilizer when their growth slows down for the winter. In temperate climates, fertilization can end in early fall and begin again, depending on the weather, in late winter or early spring. When you start to see new growth, you can safely start fertilizing. It will be different in every situation.

Pollinators

Pollinators are necessary for some plants to make fruit and seeds. The common pollinators we think of are bees and butterflies, but there are so many more animals and insects that can pollinate flowers, especially in tropical areas where most of our houseplants come from. These include bats and moths for plants that exclusively bloom at night. Hummingbirds, along with other types of birds and beetles, can be pollinators. For those flowers that smell like rotting meat, flies do the job. Flowers have ways of attracting the pollinator that works best for their shape and size. The night-blooming epiphyllum has a huge flower with a delicious smell because its main pollinator is a bat. Tiny tubular flowers are attractive to tiny hummingbirds. Of course, our plants aren't getting pollinated in our homes, but it is interesting to think about how our plants live in their natural habitats and the other forms of life they interact with and are dependent upon.

ROOT SYSTEMS AND DORMANCY

Another tip that has been widely dispersed to the houseplant community is that keeping your plant snug in the pot or rootbound is necessary for some plants to bloom. While it may be conducive to flowering, a rootbound plant is blooming because it is under stress. A plant that feels its very existence is being threatened is going to utilize all its resources to ensure it will live on in another generation. It will expend much of its energy to flower and make seeds to achieve the goal of having offspring.

This root ball indicates that the plant could use a bit bigger pot, but it is just right for a flowering plant.

I have a hoya in a smaller pot, and it does bloom regularly, but I believe it has finally reached the level of maturity needed to bloom. If your plant isn't blooming, it may be that it hasn't reached adulthood (see sidebar on page 46). A small hoya may be blooming or bloom shortly after purchase because it was propagated from a mature plant, and although small, it is really a mature plant able to bloom with all other conditions being met. Another theory explaining why a rootbound plant finally blooms is that it has been expending its energy growing leaves and roots to fill the container and once it reaches that goal, more of its energy can then be used to form flowers. A blooming plant uses a large expenditure of energy. Often plants such as bromeliads bloom and then slowly die knowing they not only have made seeds to help continue their family, but they also sent out pups (baby plants) from their base to be doubly sure. Though many plants bloom repeatedly for years while others perish after flowering once, they all still need to be at a stage in their development that is conducive to flowering.

What Is Dormancy?

Dormancy (a mandatory resting period) is another concept that comes into the conversation when discussing why a plant isn't blooming. Signs of dormancy can range from the top of the plant completely dying down to just a period of slow growth. In nature, the time of dormancy is usually concurrent with a time of hostile conditions in that plant's habitat. It may be that due to a period of excessive heat, drought, or coldness, the plants living there have developed a way to survive and withstand harsh conditions. After a long period of drought, rain results in desert plants bursting into bloom. As stated before, it takes a large expenditure of energy from a plant to flower. If it has been in a period of dormancy and the environment becomes favorable, a plant is going to take advantage of that renewal of energy and produce flowers. Remember, a plant's goal is to make more of itself by flowering, attracting pollinators, getting pollinated, and making seeds.

You may be wondering how all this pertains to houseplants. It may be that your cactus is being kept too warm in the winter for it to bloom in the spring. You may be keeping your plant too moist at a time when it needs a bit of a dry period. Not completely dry, as that may negatively affect your plant, but just a bit drier for a period. Make sure that if you are trying to get a long-night (short-day) plant to bloom, you aren't growing it in your living room where a lamp is turned on in the evening. Knowing where your plant is from and the environment it naturally lives in can give you a clue as to what it needs to flower.

Maturity matters

One major item to be noted: A plant is not going to bloom despite the conditions it is being raised in, the extreme tactics used to inveigle it to bloom, or the time of year, if it isn't a mature plant. A flowering plant will begin to bloom only when it is ready to bloom, even if proper growing conditions are met. You will find that plants grown from seed, such a citrus, for example, will not produce blooms or fruit for many years because they are juvenile plants. Yet if the plant were to be vegetatively propagated from a cutting taken from a mature plant, it could flower even on that small cutting because it is mature tissue.

PLANT HEALTH

The first line of defense in raising a healthy flowering houseplant is buying a healthy flowering houseplant. Purchase your plant from a reputable seller, whether it be an independent garden center or an online plant seller. Checking the reviews for online sellers is important because you cannot inspect the plant before you purchase. Don't head for the clearance rack for a plant, especially if you are a new plant parent. Why start with a problem from day one? If you see insects on a plant in a garden center, walk away and don't purchase plants in the vicinity. There may not be visible insects on the plant you are purchasing, but you have no idea if an egg (or many eggs) has been deposited on your plant because it was residing in the same growing space.

Make sure your plant is sleeved before leaving the garden center.

If it is cold or excessively windy, make sure the plant is sleeved, preferably in paper, before leaving the store. If it is frigidly cold outside, not only should you have your plant sleeved, but your car should be warmed, and all errands previously run so you can get your plant home as soon as possible. Nothing is worse than a plant sitting in a cold car while you run errands. The damage may not reveal itself immediately upon arriving home and unwrapping your plant. When the cold temperature damage does show itself, it will appear as limp leaves that don't recover or a mushy plant if it has been frozen. The same can be said for leaving your plant in a hot car in the summer. Neither is recommended. If you are buying a large plant that does not fit in your vehicle, hanging it out the window or standing it up in the back of a truck is a recipe for plant death. A wind-whipped plant will be damaged and may not recover. Arrange to have it delivered or rent a vehicle so the plant can lay down and be wrapped to protect it from the wind.

A healthy, well-grown plant is more likely to bloom in your home. If your plant is struggling from inconsistent care, lack of light, or an insect or mite infestation, it is going to use its energy to survive, certainly not to send out flowers. If it does manage to send out a flower when it is unhealthy, it is because it feels it may be dying and needs to attempt to make seeds for the survival of its species.

Watch for Problems

We touched on the points earlier that are necessary to keep your plant healthy. Even if all those needs are met and your plant seems to be living its best life, unfortunately, problems can arise. This usually comes from the appearance of a pest, disease, or fungus. The first line of defense is early detection. Every time you interact with your houseplant, it is important to check it carefully for early signs of a problem. Remember, a yellowing leaf isn't always an indication of a problem. Older leaves will die and fall off. This is a natural occurrence with all plants as they grow and mature. Notice the color and size of the leaves or any distorted growth that would indicate something is sucking the juice out of the leaves. Always examine underneath the leaves for insects as well as in the axils of the leaves where they meet the stem. Insects want to hide from detection, and these are places they won't be seen immediately. A magnifying glass may be helpful to ferret out some of the smaller pests, such as spider mites and thrips.

The first clue that there could be a problem is the presence of a shiny, sticky substance on the leaves, container, or floor. This is called honeydew—the excretion from insects that are sucking the juices out of the plant.

Insecticide use tips

When an insecticide or any remedy is used on plants, the label needs to be read and followed to the letter. Make sure what you are using is labeled for use on houseplants and test the product on one leaf to make sure it doesn't damage your plant before using it on the entire plant. Always start with the least toxic remedy, such as mechanical removal of the insect or spider mite. Often a stiff spray of water will remove many of the offenders. Next, try diluted rubbing alcohol on a cotton swab, insecticidal soap, or Neem oil, working your way up to an insecticide if necessary. Whichever remedy you choose to use, it will need to be repeated as necessary. The first application may not affect all the mature adults or unhatched eggs.

Spider mites are tiny and may be missed, but seeing webbing present is a good indicator.

Mealybugs resemble tiny cotton balls, and that covering is protecting the insect underneath.

Small brown bumps on the stems or leaves of the plant may be scale insects.

Powdery mildew can cover the leaves and flowers and lead to plant death.

Aphids can be many colors but prefer the newest growth, so look for them on the tips of the plants.

Armored scale are usually light colored and do not produce any honeydew.

Indoor Plant Pests and Their Control

PEST NAME	DESCRIPTION	FLOWERING PLANTS ATTACKED	PHYSICAL REMEDY	PRODUCT REMEDY
Aphids	Aphids are small pear-shaped insects (with or without wings) most often found in clusters on the new growth of plants. They can be red, black, green, yellow, gray, or transparent.	Look for aphids on the new growth of most houseplants. Though they are not often found inside, they may be seen on African violets, *Aeschynanthus*, *Streptocarpus*, and *Aphelandra*.	Wipe the aphids off with your hands and squish them. Remove them with a strong stream of water from the sink or a shower sprayer. Do not use a high nitrogen fertilizer, which will avoid having a flush of new growth.	Usually you will not have to resort to insecticides, but you may need a systemic insecticide if the aphids are not controlled. Try insecticidal soap first.
Fungus Gnats	Fungus gnats are tiny black flying insects that fly around your plants or end up in your cup of coffee or tea. The immature stage is a small larva that lives in the top 1" (2.5 cm) of the potting medium.	Fungus gnats can be found around any houseplant that is kept too wet.	To control fungus gnats, let plants dry out more between waterings, use sticky traps, and replace the top 1–2" (2.5–5 cm) of potting medium.	A last resort to control fungus gnats is to use a systemic insecticide.
Mealybugs (Foliar-Feeding)	Foliar-feeding mealybugs are small insects with a white, waxy, fuzzy-looking covering. Masses of mealybugs can be found in the axils or on the undersides of leaves. They secrete honeydew on the plant, which in turn may grow black sooty mold.	Mealybugs can be found on most houseplants, including *Clivia*, *Hoya*, succulents, cacti, and African violets.	Wipe mealybugs off with a cotton swab dipped in rubbing alcohol.	Use Neem oil, an all-season horticultural oil, or insecticidal soap.
Mealybugs (Root)	Root mealybugs are white, ricelike insects in the soil of the plant. They produce a sticky residue on the inside of the container. They are not obvious, but if the plant is failing and nothing can be found on the top of the plant, look on the roots and in the potting medium.	Root mealybugs are found most often on cactus and other succulent roots and in the potting medium.	The first task is to change all the soil.	Since root mealybugs are in the roots, they are a challenge to control. Use a systemic insecticide as a last resort.

Indoor Plant Pests and Their Control

PEST NAME	DESCRIPTION	FLOWERING PLANTS ATTACKED	PHYSICAL REMEDY	PRODUCT REMEDY
Mites (Cyclamen and Broad)	Cyclamen and broad mites are even smaller than spider mites and cause tight plant centers, brittle, curled leaves, and hairier foliage.	Cyclamen and broad mites frequently attack *Cyclamen* and African violets.	Since cyclamen and broad mites are very difficult to control, discard the plant.	If you want to attempt to control them, use a miticide.
Scale (Armored or Hard)	Armored or hard scale are usually light colored, produce no honeydew, and are hard to remove from the plant.	Hard scale often attacks cacti and other succulents, *Adenium*, bromeliads, citrus, and *Gardenia*.	Scrape off the scale with a soft cloth or your fingernail.	To control armored or hard scale use an insecticide, Neem oil, or horticultural oil.
Scale (Soft)	Soft scale are small, brown, oval bumps found on stems and leaves. They excrete honeydew, which is a good indicator they are present. Soot mold may grow on the honeydew.	Scale can be found on *Gardenia*, *Medinilla*, citrus, *Anthurium*, *Epiphyllum*, *Adenium*, and many other houseplants.	Carefully scrape off the scale. Use an alcohol-dipped cotton swab to wipe off the pest.	Use Neem oil, an all-season horticultural oil, or a systemic insecticide as a last resort to control scale.
Spider Mites	Spider mites produce gritty leaves, webbing between the leaves, and stippled or yellowing foliage.	These mites flourish when any plant is allowed to dry out and humidity levels are low. They are often found on thin-leaved plants.	Wash the foliage, or at the very least, water the plant in the shower or sink, elevating the humidity.	Control spider mites with Neem oil, horticultural oil, or a miticide (not an insecticide).
Thrips	Thrips are minute insects. Signs of infestation include spilled pollen, dried up flower petals, and damaged leaves.	Thrips can attack any flowering plant but are often found on the flowers of African violets and other gesneriads, poinsettia, and *Cyclamen*.	Remove the flowers and wipe down the leaves.	Use insecticidal soap, horticultural oil, Neem oil, or a systemic insecticide to control thrips.
Whiteflies	Whiteflies are white, flying insects that exude honeydew and cause yellowing and stippling of the leaves.	Whiteflies can attack any leafy, flowering plant. They are frequently attracted to *Gardenia*, *Clerodendrum*, citrus, begonia, *Spathiphyllum*, and poinsettia.	Whiteflies can be controlled by vacuuming the insects or using yellow sticky cards.	Control whiteflies with an indoor plant insecticide.

Flower Portraits

In this chapter, be prepared to add some plants to your must-have list. Flowering plants add so much beauty to our homes and are fun to grow. There is much pride to be had when your plant bursts into bloom and you know you had a part in it. The chapter is divided into groups by plant family, including a section about holiday plants and more blooms that didn't fit into any other category.

First, we begin with the bromeliads, a family with not only unique blooms but beautiful foliage. Next, cacti and other succulents take the stage. These often misunderstood plants can be grown in our homes with some specialized care and can even be coaxed into flowering. The next group of plants are probably my favorite because the gesneriad family includes the African violet. The other family members are often overlooked, but there are so many beautiful, easy-to-grow cousins that should be in the spotlight. The orchid family needs no introduction. These flowers have mesmerized collectors for decades. The holiday plants are those plants known to all of us as flowers we give, receive, or decorate with during the holiday season. Finally, the plants that didn't fit into a group but needed to be recognized for the beauty they could add to our homes. I hope you find a plant (or many) you think you would like to add to your collection. Flowering plants may need different and more specialized care than foliage plants, but the extra attention will be worth it when your plant bursts into bloom!

Aechmea fasciata

Silver Urn

This common bromeliad is easy to grow and is widely accessible to consumers. Silver urn is a fascinating bromeliad that was given the name *Aechmea* by Spanish botanists Ruiz and Pavón in 1794 from the Greek word *aichme* meaning "lance head" or "spear-head," referring to the sharp points on the flower bracts and sepals. This plant also has sharp spines on its leaves and greatly resembles its cousin, the common pineapple. It is a "tank" bromeliad growing in a vase shape, and the tank must be filled with water. In nature, these tanks often have organisms living in them, such as small frogs, and when small insects, plant debris, and bird droppings fall in the tanks, they decompose, supplying the plant with nutrients. In our homes, you may want to add a much-diluted fertilizer to the water occasionally, as it won't have natural fertilization.

LIGHT: To support this plant and bring it into bloom, place it in a bright light. Though it may survive in low light conditions, it won't be as colorful and its leaves may become thin and floppy whereas in better light, they will be stiff and keep their vase shape. An east or west window would be preferred as direct south sun may cause sunburn on the leaves. If you don't have enough window light, growing your *Aechmea* under electric lights will work fine.

WATERING: Keeping the tank filled with fresh water is imperative. When you water, pour it in the tank until it overflows, wetting the potting medium. If you buy this plant and it's residing in a plastic grower pot, it would be best to move it into a similar size clay pot or some other heavy pot, as this plant can become top heavy, especially when it sends out its large flower stalk.

FLOWERS: Though the most colorful part of the flower stalk is its pink bracts, the small bluish flowers are also attractive, appearing as the flower spike expands out of the tank. The flowers are replaced by berries that extend the season of beauty. After blooming, the mother plant will slowly start to expire, during which it will send out new babies, or "pups," at the base. Those pups are her replacements and will furnish you with more plants.

If you have a mature plant that refuses to bloom and is in plenty of light, it can be forced to bloom. Use an apple, which, as it ripens, releases ethylene gas (a natural plant growth regulator), and place it in a plastic bag with the plant, emptying the tank first so the plant does not rot in the bag. After approximately a week, remove the bag, throw away the apple, and refill the tank with water. The plant should bloom after this treatment.

PROPAGATION: When the pups are one-third the size of the mother plant, they can be cut away and potted in their own containers. Seeds can also be used to start new plants.

TOXICITY: This plant is not toxic to cats or dogs, but it is doubtful they would chew on it as it has barbed leaf edges.

Billbergia species

The most common *Billbergia* is the Queen's tears, or *Billbergia nutans*. When you first see this plant in bloom, you will be amazed. The beautiful, rich pink- and blue-colored flowers drape down from the narrow silver foliage. Other varieties of *Billbergia* have flowers that are less weeping and more upright yet equally as beautiful. The plant itself is a tall rosette of leaves that together form a "tank" or vase, which should always be filled with water. Not only do they have beautiful flowers, but the foliage can range from plain silver to burgundy, and some may even have colorful stripes or dots on the foliage. When *Billbergia* aren't blooming, they are still show-stopping plants because of the foliage. *Billbergia* are named after Swedish botanist Gustaf Johan Billberg (1772–1844), and their native habitat is in Brazil, Paraguay, Uruguay, and northern Argentina, where they grow primarily as epiphytes attached to other plants. They don't have extensive root systems and use their roots to attach themselves to other plants. In our homes, the roots do help keep the plant in their containers. But remember that water in the vase is more important than having water around the roots. Because *Billbergia* don't need an overly large container, use one that is heavier, such as clay or glazed pottery, so that these top-heavy plants don't tip over.

LIGHT: *Billbergia* need bright light to bloom and thrive. They don't need to be in the full sun to keep their coloring and rigid shape, but if they are in too little light, they will become less upright and the softer leaves may droop.

WATERING: This is an easy plant when it comes to watering. Simply fill the tank or cup in the middle of the plant. Add water until it runs out of the tank and spills into the potting medium. Changing the water in the tank regularly helps keep disease and rot at bay.

FLOWERS: The scape or flower stalk arises from the middle of the plant out of the tank and puts on its display above the foliage. The bracts or modified leaves may be more colorful than the actual flowers, but together they are lovely.

PROPAGATION: Propagation can be achieved by splitting the pups or babies away from the mother plant when they are one-third her size. Pot them up in individual small pots. You can even leave the babies in the original pot, allowing for a larger display of plants. *Billbergia* can also be started from seed, but it is better to split off the pups, as they will flower sooner.

TOXICITY: This plant is not toxic to cats and dogs, but I doubt they would choose to nibble on the leaves that have barbs along the edges.

Guzmania lingulata Hybrids

Scarlet Star

This bromeliad from the jungles of southern Mexico and Central and South America may not always have scarlet bracts despite its common name. They may be orange, yellow, red, purple, pink, or any shade between. The colorful part of the plant is the bracts or modified leaves, while the actual flowers are small and resemble white matchstick heads that rise above the bracts when they are flowering. Since the flowers are fleeting, it is the color of the bracts, which lasts for months, that gives the plant interest. *Guzmania* has smooth, thin leaves that are much different than many other bromeliads with their stiff leaves and sharp spines. This plant is usually epiphytic, growing in the shaded undergrowth rather than in the tops of trees where the sun is blazing. This accounts for the thinner, greener leaves that are present on these plants. This is another bromeliad plant discovered by explorers Ruiz and Pavón and named after Anastasio Guzman, an eighteenth-century Spanish apothecary.

These plants are often used as underplantings in interior landscapes because they provide long-lasting, affordable color. When the color fades, they are replaced with another flowering plant. These plants are also used outside as bedding plants in the southern United States like annuals are used in the northern states. There are also variegated forms of *Guzmania* with stripes or white outlined leaves.

LIGHT: *Guzmania* need a bright light to form their colorful bracts, but not direct sun as that may burn their thin leaves. Plants that are typically epiphytes in their native habitat under the canopy of trees usually make great houseplants as they naturally survive in lower light levels. *Guzmania* will do well in an east or west window, or they could be grown under electric lights.

WATERING: This bromeliad is one of the tank or vase types that collect water in the receptacles in their native habitat. Therefore, in our homes, that tank must be kept filled with fresh water. Change it completely at least once a month to deter fungi or diseases from taking hold. In their native habitats, they have the benefit of good air circulation to keep fungus at bay and lots of rainwater to frequently rinse out the tank.

FLOWERS: As mentioned above, the *Guzmania* flowers are not as decorative as the bracts that surround them. These bracts appear in mostly "warm" colors on the color wheel, including red, yellow, orange, and pink. The bracts of *Guzmania* last the longest of all the bromeliads. After these bracts fade, the plant will send out pups that, after attaining one-third of the size of the mother plant, can be cut away and potted up separately. They can be left to become a clump of plants, too.

PROPAGATION: Like all bromeliads, after the plant blooms, it will slowly die, but not before sending out pups or babies at the base of the plant. *Guzmania* can also be propagated by starting seeds.

TOXICITY: This plant is not toxic to cats or dogs.

Tillandsia ionantha

Air Plant

Air plants (*Tillandsia*) are fascinating plants because they aren't typically grown in containers, and they appear to be growing with no sustenance but the air. If you live in the southern part of the United States, the trees are dripping with large clumps of Spanish moss. Many don't realize it, but this is a living plant, and it is a tillandsia (*Tillandsia usneoides*). *Tillandsia* were named after the Swedish botanist Elias Tillandz. It has been cited that Mr. Tillandz had a debilitating fear of water and would go around a lake rather than cross it. It's a good thing tillandsias don't grow in water.

Other types of bromeliads such as *Neoregelia* and *Aechmea* also grow on trees like the *Tillandsia* and are close relatives of the pineapple (which is a terrestrial bromeliad). These plants that are growing on other living plants can appear to be hurting their host plant or taking nutrients from them. Yet, they are not parasitic; they are simply using the plants as a place to attach themselves and grow.

How do they survive without living in soil or receiving their nutrients from the trees? *Tillandsia* have special water-collecting receptacles called trichomes covering them entirely. They collect water and store it until the plant needs it. In nature, they receive their nutrients from the rain. Their roots aren't taking up water and nutrients like most plants but are merely holding the air plant to whatever it is clinging to. Though we see them in the south clinging to other plants, you will also find them clinging to utility wires, rocks, and other nonliving objects.

(continued)

This *Tillandsia ionantha* is done blooming, but the color of the leaves remains for a short time afterward.

Tillandsia caput-medusae has a silver sheen on its leaves from all the water collecting trichomes.

Air plants are easy plants to take care of in your home. Problems arise when people assume since they aren't attached to anything, they can be placed anywhere in the house. They are often used as décor, placed in a bowl on a coffee table or on a shelf on the wall. This is fine if you are providing the light they need to thrive.

LIGHT: *Tillandsia ionantha* do best with as much light as you can give them. They grow well in west and south windows, blooming regularly and making more pups. Hanging them in glass orbs near the windows helps keep the humidity higher and allows them to be closer to the light. A common mistake is placing them in too low light conditions where they often rot because they can't use all the water they have stored. The more silvery the plants, such as *Tillandsia xerographica* and *Tillandsia tectorum*, the more they need higher light levels and less water than their greener, smoother leaved counterparts such as *Tillandsia brachycaulos*. The large amount of trichomes these specific plants have allow them to store more water for future use. They are often found growing in drier tropical areas with little rainfall.

WATERING: When watering your *Tillandsia*, it is best to submerge them completely in a sink or bowl of water. Usually, your typical tap water isn't a problem, but some people opt for using distilled, rain, or bottled water for their *Tillandsia*. Allow them to soak for thirty minutes or more, shake the excess water off, then turn them upside down to drain and dry. As they grow on trees and other objects, they are never naturally growing straight up and down, and if the water collects in the middle of the plant, they could rot. Remember, in their native environments,

Spanish moss (*Tillandsia usneoides*) hangs from trees in the southern United States. This one has a seed pod left from the flower.

Tillandsias grow as epiphytes on trees in their natural habitat as this picture, taken in Florida, USA, shows.

they typically receive nice tropical breezes to help them dry out.

Misting can be done between thorough soakings but should not be relied on exclusively to keep your plant hydrated. Fertilize with a weak solution of water-soluble fertilizer in the water. As stated above, the more silver a plant, the less water it needs, so the *T. xerographica* and *T. tectorum* won't need to be soaked as often as some of the other *Tillandsia*.

FLOWERS: Most *Tillandsia* turn blush pink or red before they are ready to bloom. They then send out thin purple flowers that last a short time, but the colorful blushing plant will stay that way for some time. Other *Tillandsia* will first send up large inflorescences that the smaller purple flowers emerge from. After the flowers are done, the inflorescence can be removed.

PROPAGATION: After blooming, the plant begins to die, but not before sending out pups or baby plants. There are usually two new plants to take the place of the one that will expire. The mother plant doesn't die immediately, and if you allow the babies to continue to grow and flower without splitting them apart, eventually you will have a large multiplant cluster. If you prefer, when the baby is one-third to one-half the size of the mother plant, you can carefully pull it from the mother plant and grow it individually.

TOXICITY: This plant is not toxic to pets. I have found that my cat likes to bat air plants around the floor like a mouse, so they eventually fall apart, but he doesn't seem interested in eating them.

Trichomes

Bromeliads are covered with water-collecting receptacles called scales or trichomes. Since bromeliads most often are epiphytes, these plants aren't collecting water through their root systems as their roots are primarily used to adhere the plant to the surface of what it is living on. The more silver the bromeliad, the more trichomes it has. Often bromeliads that are living in more xeric areas (*Tillandsia xerographica* and *Tillandsia tectorum*) will have more trichomes as they must reserve water longer than other bromeliads that live in more moisture-laden regions. They also serve to protect the leaf surface from solar radiation at higher altitudes.

The silver covering on this tillandsia are trichomes, which store water.

Wallisia cyanea (formerly *Tillandsia*)

Pink Quill, Blue Flowered Torch

If you were to find one bromeliad consistently offered for sale in most garden centers and big box stores, it would be the pink quill—so named for its pink inflorescence, which resembles an upright feather. Its Latin species name *cyanea* means "blue" and describes the color of its flowers. The purplish-blue flowers consist of three large petals and emerge from the crevices of the bracts. The vibrant flower color pops against the bright pink bracts, making a beautiful display. After the flowers have faded, the pink bracts will remain colorful for months.

Pink quill was recently changed from the genus *Tillandsia* to the genus *Wallisia* after DNA analysis was performed on the plant. As it was previously in the *Tillandsia* or air plant genus, you may think it needs no soil, but this plant is always offered for sale in a container. Its root system gives the plant water and nutrients, unlike *Tillandsia*, which receives moisture and nourishment from its leaves.

Pink quill hails from Ecuador and Peru, where it grows as a lithophyte on rocks or as an epiphyte on tree branches. Unfortunately, it is believed to be extinct in the wild from over-collecting and habitat destruction. After it blooms, the inflorescence will die back as new pups or babies grow from the base of the original plant.

LIGHT: To ensure the plant flowers, it needs to be in a bright light. Place your plant where it will receive some direct sun during the day, such as an east or west window. You may choose to grow your *Wallisia* under lights, leaving the lights on for seven to fourteen hours per day, depending on whether fluorescent or LED lights are being used.

WATERING: Water the potting medium with tepid water, making sure it drains through the bottom of the container. Use a well-drained potting mix, adding some orchid bark to make a chunkier mix that will allow plenty of oxygen to surround the roots. As it grows naturally as an epiphyte, *Wallisia* is used to having plenty of air circulation around the plant and the roots, which is important for its health.

FLOWERS: The flowers appear up the sides of the pink inflorescence, emerging from the crevices where the overlapping bracts meet. They are a purplish-blue, three-lobed flower that only last two or three days, but the pink bracts may last a few months. The shocking pink bracts are the reason most people buy this plant as they last a long time, giving months of color.

PROPAGATION: After blooming, the mother will send up pups or baby plants from its base as she slowly expires. When the pups are one-third the size of the mother plant, they can be separated into their own containers. If you would like a larger, fuller container of plants, leave them to grow in the same pot, moving them to a larger container when needed.

TOXICITY: *Wallisia* is not toxic to cats or dogs.

Adenium obesum

Desert Rose, Mock Azalea

The desert rose has become easier to find in the last few years and is often sold as a small bonsai. Thankfully, the price has come down as they become more widely available. Its large caudex or swollen stem makes it a popular plant because it is often quite gnarly looking. The caudex is a water storage facility as *Adenium* are used to drought in their native desert habitat of Africa. Most of its flowers are on the tips of the branches, and the plant stems can look quite sparse, yet when the blooms emerge, all sparseness is forgotten. Their trumpet-shaped flowers are stunning in shades of red, white, pink, and burgundy—often with splotches and stripes of color. This plant has a shallow and wide root system, so a low bowl planter is preferred. As it needs good drainage and does not want to remain too wet, porous clay pots are recommended.

LIGHT: Give this plant as much light as possible as it would prefer to be living in the desert in full blazing sun. Place it in an unobstructed south or west window. Lower light may be acceptable, but it won't flower as prolifically, and the stems may become elongated and soft. If you take your plants outside in the summer, this is one that would enjoy a summer sojourn in the sun. Remember to slowly acclimate your plant to the sun outside and to the lower light when it is brought back in for the winter. Make sure the temperature doesn't drop below 40°F (4°C) before you bring your *Adenium* inside.

WATERING: First, this plant needs to be potted in a fast-draining potting medium; this is a desert dweller, after all. In the summer, it can be watered liberally, especially if it is outside in the full sun. In the winter, when it is resting, give it minimal water if any at all. It is important not to water your *Adenium* if it is in a cold spot. Your plant will rot if you water when it's cold. But make sure the caudex doesn't become wrinkly, which would indicate it is too dry. As it goes dormant in the winter months, it may defoliate and that is okay.

FLOWERS: The trumpet-shaped flowers appear in the spring and summer in clusters at the tops of the stems. They range in color from white to pink to red with stripes and mottling of these colors on the flowers. They need cross-pollination to set seed and are pollinated by hummingbirds, butterflies, and bees.

PROPAGATION: *Adenium* can be propagated from stem cuttings but will produce the swollen stem or caudex faster if started from seed. If you take a cutting, remember the sap is poisonous and handle with much care, gloves being strongly suggested. Allow the cutting to callus over before placing in a pot of moist potting medium.

TOXICITY: This plant is extremely poisonous and should be kept away from children and pets. Its sap has been used to make poisonous arrows.

Ceropegia species

String of Hearts, Rosary Vine, Hearts Entangled, Sweetheart Vine

These endearing, often diminutive flowering vines are right up there in popularity with the other "string of" plants, such as string of pearls, turtles, and dolphins. The string of hearts or rosary vine, *Ceropegia woodii* is an older variety that has been cultivated for the houseplant industry for years and has lately come back in vogue—especially the variegated form. The flowers can be easily missed because the leaves also have a purplish cast, especially when they are receiving plenty of light. These long vines can also be trained around a topiary frame or up a trellis. It would be best to start training the plants when young because the hearts entangled moniker will become clear to you when you try to untangle long vines to wrap them around a frame. When this happens, it needs to be done delicately and slowly to ensure minimal loss of leaves.

Ceropegia was named by Carl Linnaeus. It was believed that the flowers resembled fountains of wax, and the plant was named from the Greek words *keros* meaning "wax" and *pege* meaning "fountain." The species name *woodii* was used to honor English plant collector, John Medley Wood (1827–1915).

There are other lesser-known varieties that don't have the endearing mottled heart-shaped leaves, but they do have larger, hard-to-miss flowers. The parachute plant or umbrella vine, *Ceropegia sandersonii*, is a variety worth seeking out. Its large greenish flowers do resemble an opened parachute. This *Ceropegia* is considered a semi-carnivorous plant as it gives off a scent that lures flies to its flowers, but the fly is not killed. The unaware insect climbs into the flower, and downward facing hairs prevent it from crawling back out until the flower begins to wither and the hairs collapse, allowing the fly to escape. Unfortunately for the fly, but fortunate for the plant, no lesson is learned, and it flies to the next *Ceropegia* flower and repeats its mistake while covered with pollen from the last flower. Its pollinating job is complete. Another variety if you are collecting the "string of" plants is the string of needles or *Ceropegia linearis*. Its leaves are thinner and pointed at the end, without the heart shape of its cousin.

LIGHT: These flowering vines need bright light but don't require full sun. Place them in a west or east window, or back 1 to 2 feet (30 to 60 cm) from a south window. If they are in too low light, the internode (stem length between the leaves) will elongate, giving the plant a stragglier appearance. If it is grown in the right amount of light, the internodes will be shorter, giving the appearance of a fuller plant. If it doesn't flower, move it closer to a light source.

(continued)

LEFT: *Ceropegia ampliata* has a common name of bushman's pipe.

Ceropegia sandersonii has a common name of parachute plant.

The heart-shaped leaves of the *Ceropegia woodii* is the characteristic that draws people to this plant.

WATERING: During the summer months when this plant is actively growing, give it thorough waterings when the soil has almost completely dried out. In the winter when it isn't actively growing, an occasional watering will be sufficient to keep the leaves from shriveling. If it is allowed to become that dry, the stems may die back where they spill over the pot. When too dry, the succulent stem will flatten on the pot rim and may not be able to recover when sufficient water is applied. If that happens, cut those stems off and use the tips to propagate new plants. Usually, these vines scrabble over other vegetation in their natural habitats so the stems are supported by other plants. Also, when they get extraordinarily long, the stems can become too heavy to support the weight.

Ceropegia woodii 'String of Arrows' in flower.

FLOWERS: The flowers on these *Ceropegia* vines are interesting and weird. The base of the flower is usually bulbous in nature, which adds to the uniqueness of it. The tip of the flower is different from species to species. The string of hearts and string of needles have small purple pipe-shaped flowers with dark purple, fuzzy-looking petals that are fused together at the top. *C. sandersonii* flowers are green with white hairs around the rim.

PROPAGATION: *Ceropegia* are easy to root from stem cuttings. Lay the cuttings on top of a pot of moistened potting medium and the roots will emerge from the cut end and grow into the soil. It can also be propagated from the small tubers, which resemble tiny potatoes, that occur on the stems of the plant. Cut them off with a piece of the stem, and as before, lay them on top of a pot of moist potting medium and roots will begin to grow. Use a few cuttings in a pot so that it grows fuller faster. Use a gritty, well-drained cacti potting medium.

TOXICITY: *Ceropegia* is not toxic to pets or humans.

Cleistocactus winteri

Golden Rat Tail Cactus, Golden Monkey Tail, Tarantula Cactus

I was given a specimen of this plant many years ago and was enamored with the orange flowers that appeared out of the stems in random places. The stems can be up to 1 inch (2.5 cm) in diameter and may grow up to or beyond 3 feet (91 cm) long. This unusual cactus is from the forest cliffs of Bolivia, where it cascades over the rocks looking like a giant yellow spider. It is an easy cactus to grow in our homes but needs to be placed in as much light as possible. This cactus begins its life growing straight up, but as it gets taller and heavier, it will bend down over the pot, becoming more of a cascading plant, making it well-suited for a hanging basket. Often this plant is found as a crested (cristata) form grafted to a *Hylocereus* cacti. This gnarled form has no resemblance to the long rat-tail appearance of the noncrested form other than having yellow spines. It is a popular collector's item—as many crested cacti are.

During the summer, this plant could be moved outside to soak up the summer sun. Make sure to acclimate it to this new place in the sun for a couple of weeks by placing it under a tree or on the north side of your home first. Bring it in before it freezes in the fall, acclimating it first by placing it in a shaded place so that it is better accustomed to the lower light levels in your home.

LIGHT: To promote blooming in the home, this plant will need as much light as can be provided. Close to a south window would be optimal with a west window as a second choice. Growing your cactus under electric lights will help it bloom if enough natural sunlight isn't available.

WATERING: As with most cacti and other succulents, the *Cleistocactus* stems are water-storing receptacles. Therefore, they are quite drought tolerant. Water it thoroughly and then allow it to almost completely dry out before watering it again. If it is outside for the summer, protect it from fall rains so that it isn't wet and cold as this could cause root rot and possibly the death of your cactus.

FLOWERS: The flowers appear at random spots on the stem of this cactus. The blooms can be found on mature plants from spring through the autumn at random intervals. The flowers are an orange or salmon color and last for several days.

PROPAGATION: This plant can be propagated from cuttings of the stems. The key is to allow the cuttings to callus over for a couple of days. Use a very well-drained potting medium formulated for cacti and other succulents and place the cuttings only as far into the medium as needed to stand up. Use small bamboo skewers to hold them up if needed until they are rooted.

TOXICITY: This cactus is toxic to dogs and cats, but the spines should deter them from munching on it.

Epiphyllum oxypetalum

Queen of the Night, Orchid Cactus

Queen of the night is a perfect name for this stunning orchid cactus. It hails from Mexico, Central America, and northern South America, where it grows epiphytically. Though it is called an orchid cactus, this plant is not an orchid, but as it grows in conditions comparable to some orchids and has amazing flowers, the name is understandable. This jungle cactus would naturally be seen scrambling through the treetops in dappled sunlight sending out its gorgeous flowers. They are most often offered for sale in hanging baskets as they tend to droop, especially when the large white flowers appear. This genus name comes from the Greek *epi* meaning "upon," and *phyllon* meaning "leaf," which refers to the large white flowers that emerge from the edges of the leaf. Yet, the large "leaves" are not true leaves, but flattened stems.

LIGHT: These plants need enough light to bloom, but placing them in direct sunlight in a southern exposure may result in burned areas on their stems. A western exposure would work well, or place them back 1 foot (30 cm) or so from a direct southern exposure.

WATERING: These epiphytic plants would prefer to be in a quick-draining, coarse potting mix. Remember, they are epiphytes and so grow naturally in forks of trees in the jungle canopy. In nature, they receive large amounts of rain but never stand in water as it quickly drains away. A potting mix comprised of regular potting mix, with perlite or pumice mixed in with some orchid bark, should work well. Water it well and never allow it to completely dry out. *Epiphyllum* would prefer some extra humidity as well.

FLOWERS: The flowers of this not-so-attractive plant are striking if you are awake to see them. As the common name suggests, the flowers open at night. You must pay close attention so that you don't miss the show. When the flower bud begins to curve upward and the white petals begin to emerge from the dark pink sepals, it is a good sign that the flower will open that night. As the sun dips below the horizon, these flowers wake up and the show begins. Some people have flower parties so others can witness the amazing phenomenon. *Epiphyllum oxypetalum* flowers are open all night, but as dawn approaches, they close and deflate like a popped balloon. While they are open, the flowers are breathtaking; enormous with a diameter of up to 9 inches (23 cm) and the scent they emit is intoxicating. Who are they trying to impress in the middle of the night? Certainly not humans. Their glow-in-the-dark white flowers and beautiful scent are meant to draw in the pollinators they need to help with the fertilization process. For these flowers, the primary night pollinators are bats—moths may also play a role in pollination. The key to the whole process is that the bat remembers where the flowers were and returns (along with other animals) to eat the fruit, thereby spreading the seeds around to help the plant spread.

(continued)

When the bud begins to show its white petals, be ready because the flower will open that evening.

When the sun comes up, the flower is done and closes up, hoping a pollinator has done its job.

PROPAGATION: Propagation of the orchid cactus is not difficult. Take stem cuttings approximately 4 to 6 inches (10 to 15 cm) long and lay them aside for a few days to allow the cut end to callus. Then place the cut end into a pot of the same medium used for your mother plant. It will send out roots and then new stems. Plants from cuttings should bloom in two to three years.

TOXICITY: This plant is not toxic to pets.

Epiphyllum Orchid Cactus

While the *Epiphyllum oxypetalum* is considered an orchid cactus, there are other members of the same genus that have large colorful flowers and bloom during the daytime hours, with flowers lasting more than one day. Plants that are white and bloom at night are targeting the moth and bat population to assist with pollination. Their colorful family members compete with other jungle flowers for pollinators during the day, so they must be colorful and flamboyant to win the competition. These beautiful flowers come in red, pink, orange, yellow, purple, and all shades in between and are attractive to butterflies, bees, and hummingbirds. They emerge from flattened stems on plants that are quite gangly and not so attractive when they aren't in bloom.

Orchid cacti are epiphytes in their tropical homes, scrambling through the treetops finding patches of light. Because they are living in the trees, they do not need full direct sun to bloom. In fact, if they are in sunlight that is too intense, their stems can become sunburned. Instead give them bright light with some direct sun in the morning or afternoon.

The key to getting them to bloom prolifically is to make sure they get a cooler period in the winter so they go dormant but with temperatures that don't fall below 55°F (13°C). Because they don't have extensive root systems, keeping them snug in their pot will also help with blooming.

Euphorbia milii

Crown of Thorns, Christ Plant, Christ Thorn

This Madagascar native can be continuously in bloom in the home environment. Though it appears to be a cactus because of the spiny nature of its stems, it is a succulent. Those stipular spines lack an areole as they would have if they were cacti. If you bump into one, you won't be able to tell the difference though. If you prefer non-spiky plants, search for *Euphorbia geroldii*, which is spineless but has the same crown of thorns flowers.

When a leaf is removed or the plant is cut, you will find that a sticky white sap exudes from the damaged areas. This is a latex substance, and contact with skin should be avoided—so handle crown of thorns with care. It is especially important to keep the sap away from your eyes. These plants offer year-round color on the windowsill. The plant received the common names crown of thorns or Christ plant as it is thought to have been used for the crown placed on Christ's head at his crucifixion. The genus name *Euphorbia* was used to honor Euphorbus, a Greek physician, and the species name *milii* to commemorate Baron Milius, governor of Réunion (formerly Bourbon) and responsible for introducing the plant to France.

As this plant is a succulent, a well-drained soil is imperative. Use a cactus and succulent growing mix and add a good amount of pumice, perlite, or poultry grit to aid with drainage.

Euphorbia milii comes in many sizes, from mini plants to plants that exceed 5 feet (1.5 m). The original color of the flowering structure is red, but through much hybridization, there are now colors ranging from red to yellow to white with variegated forms also available. Not only were the hybridizers trying to create more colors, but the size of the flower structures has also become much larger for a more striking display. Some of the hybrids from Thailand have flower heads so large, they resemble hydrangeas. There are cultivars with variegated foliage as well.

LIGHT: Give your crown of thorns as much light as possible for the best flowering results. An unobstructed south or west window will ensure flowers throughout the year, but growing your plant under electric lights will also result in a continuously flowering plant.

WATERING: Though these plants are succulents and store water in their stems, if they are allowed to dry out, the leaves will yellow and drop. Give it a thorough watering, meaning water is coming out of the drainage hole and the entire root ball is moistened. Allow it to almost dry out before watering thoroughly again. If it is grown in high light, keep it on the dry side but realize it may need more water than you think. If you take your plants outside for the summer, bring them back inside before night temperatures drop below 60°F (15°C). Euphorbias do not like cold temperatures and will drop their leaves quickly if chilled. The plants will most likely survive even if the temperatures fall to 40°F (4°C), but they will defoliate. If the soil is wet and the plant is cold, death may occur.

(continued)

After much hybridizing, crown of thorns now is available in numerous flower colors.

FLOWERS: The crown of thorns flower (and that of other *Euphorbia*) is called a pseudanthium. These plants have an inflorescence that resembles a flower, but the colorful parts of the plant are bracts or modified leaves. These bracts are fused together in the *Euphorbia* and called a cyathium. The actual flowers are in the middle of that cyathium and are yellowish-green. The colorful bracts attract pollinators to the plant and guide them to the area where pollination occurs. This often confuses people because the colored parts appear to be flowers. These cyathia could be from ¼ inch (6 mm) to 1½ inches (4 cm) across. As mentioned above, the Thai hybrids have cyathia that are so large that a cluster of them resemble a hydrangea blossom. Of course, these are on much larger plants as well, so make sure you have room before choosing one of those plants.

Bracts

Often what people think of as flowers on a plant are technically bracts, or modified leaves. The flowers are usually small and inconspicuous, so colorful bracts are necessary to lure pollinators to the flowers. Often the bracts will stay beautifully colored for weeks after the flowers are done blooming. The most well-known plants with colorful bracts are poinsettias, bromeliads, and crown of thorns.

The bracts and the flower together are called a cyathia.

PROPAGATION: Crown of thorns can be propagated from tip cuttings. Carefully cut a 3 to 6 inch (8 to 15 cm)-long piece of stem and allow the end to dry and callus over before placing it in a pot of moist potting medium. You can stop the latex from bleeding out by dipping the end in water. These plants tend to grow lanky, only growing leaves on the tips of the branches. Many have been hybridized to have better branching habits, so they remain more compact. If your plant is getting leggy, cut it back and use the ends of the stems to start new plants, even adding them to the pot your plant is already growing in for a better display. If your plant has become super tall and straggly, you may want to start new plants from the tips and discard the original plant.

TOXICITY: This plant is toxic to pets and humans.

Hoya species

Wax Plant, Porcelain Flower, Wax Vine

Hoya boast beautiful, often aromatic flowers, yet until recently they were seldom found to buy. If found, it was usually the common *Hoya carnosa* or Hindu rope plant, *Hoya carnosa* 'Compacta'. Today there are numerous types on the market. Why the sudden interest? If you've ever seen a wax plant flowering, you will understand why. Often the scent that is present when it flowers is amazing, ranging from sweet to spicy with many variations in between. Add to this the fact that the foliage of many *Hoya* is remarkable with splotches and variegation that is like no other. They also come in sizes ranging from tiny to extremely large. Some collectors like to put them in more sun than necessary to "sun stress" them so the leaves turn shades of pink to red. *Hoya* are often epiphytes in their natural habitat in the countries of Asia and Australia but can also grow on rocks and scramble along the ground.

LIGHT: *Hoya* need bright light to bloom well. An unobstructed south or west window will give your plant enough light to bloom (and they may even flower in an east window). Too little light may keep your hoya green and living, but you will probably never witness their exceptionally beautiful flowers. Since there is such a large variety of *Hoya* with different needs, try yours in different windows to see how they do. They can also be grown under electric lights if your home doesn't have enough light to support good growth and flowering.

WATERING: *Hoya* prefer a well-drained soil and though most are succulents, if they are allowed to dry out too much, they will develop yellowing leaves. A cactus and succulent potting medium with extra perlite, pumice, orchid bark, or grit added would be best. Water your *Hoya* thoroughly, allowing it to dry out quite a bit before watering again. If you have *Hoya multiflora*, the shooting star hoya that has thinner leaves that are more like a regular houseplant than a succulent, pay closer attention to the water levels as it may need water more often than the succulent types.

(continued)

This *Hoya pubicalyx* bloom has a lovely aroma as well as being beautiful.

Hoya obovata flowers.

Hoya multiflora, or the shooting star hoya, does not have succulent leaves like most hoyas.

FLOWERS: The flowers of *Hoya* are greatly diverse. Their flower clusters are umbel shaped and often resemble a flower-covered, upside-down umbrella. The individual flowers are star shaped and can be waxy or fuzzy looking with fringed edges. The flower colors range from white to green to pink and yellow, and the centers are a different color for a beautiful contrast. The amount of nectar exuding from the flowers can be excessive, so when your plants bloom, make sure they aren't hanging over a nice piece of furniture that could be damaged from dripping nectar. The flowers emerge from a small spur called a peduncle. When the flowers fall, don't cut that small spur off as the flower will emerge from there again every time it flowers. If removed, it may take a while for another one to grow. The aroma of the flowers can be intense, and though some may not smell good, most smell amazing and the strongest at night. If you have a window in your bedroom that provides the right light, a nice smelling *Hoya* would be perfect.

It has been said that a snug pot helps with blooming, and I find that to be true. If your plant has been in the same pot for many years though, a container with some fresh potting medium would be better for the health of the plant. It doesn't necessarily need to go in a larger container; just give it fresh potting medium.

The flower buds of the *Hoya* emerge from a peduncle that should not be cut off after the flowers fall.

PROPAGATION: If you've seen the single heart-shaped leaves of *Hoya kerrii* sold around Valentine's Day, you may have been duped into buying one thinking it would become a large plant. What you may not know is that to propagate *Hoya*, a piece of stem (preferably with a couple of leaves) is necessary for it to grow into the vine it is meant to be. A single leaf without any stem tissue will grow roots but will stay in that form without growing into a vine.

To get a proper vining plant, take a cutting of the vine with at least two leaves on it, let it callus over for a few days, and then place it in a pot of moist potting medium or sphagnum moss. Patience is needed, but a new plant will begin to grow in a few weeks' time. You can also root them in water, allowing the roots to grow to approximately 1 to 2 inches (2.5 to 5 cm) long and then pot them into the potting mix of your choice. Use a pot that isn't too large for your plant as hoyas prefer to be snug in their pot and they usually won't bloom until the roots fill the pot.

TOXICITY: *Hoya* is not toxic to pets.

Huernia zebrina

Lifesaver Plant, Little Owl Eyes

If you are looking for a new succulent for your windowsill, you don't need to look any further than the genus *Huernia*. These African plants were discovered by and named after Justus van Heurne (1587–1653), a Dutch missionary, botanist, and doctor. The genus was misspelled but still honors the missionary. These succulents have small spikes up the sides of the four- to six-sided stems. The spikes though are soft, not sharp, like most cacti. The plants aren't much to look at, but the flowers are the stars of the show. *Huernia* have shallow root systems, so grow them in low bowls or clay pots.

Since the popularity of fairy gardening has grown, these plants are much easier to find. Look in the fairy gardening section in the garden center for many unusual small plant varieties. The small stature of these plants makes them perfect for a windowsill. While you are looking, you may find *Huernia schneideriana* or red dragon plant as well. This plant resembles *H. zebrina* but has burgundy star-shaped flowers without a "lifesaver" in the middle. Both make wonderful, easy houseplants.

LIGHT: Though most succulents have high light requirements, *Huernia* do well in medium to bright light so are easier for northern indoor gardeners to have success with. If they are in too high light, they may pale in color or even burn. They will bloom in an east or west window if set close to the windowsill. Mine do well in a south window but are a bit paler than normal.

WATERING: Use a potting medium that drains quickly, such as a cactus or succulent mix, so your plant doesn't stay too wet. Add a large proportion of an amendment such as pumice, grit, or perlite. Make sure your container has a drainage hole so water is not left standing in the bottom of the pot. Thoroughly water until water runs out the drainage hole, emptying any excess in the saucer. Allow the medium to dry out almost completely before watering again.

FLOWERS: The flowers of *Huernia zebrina* are not large but definitely are unusual. The flower is star shaped with a center ring that resembles a rubber lifesaver or donut. The flower is a yellow-golden color with dark brown stripes, and the "lifesaver" is burgundy with the whole flower measuring about 1 inch (2.5 cm). I don't find they smell bad, but some people do. In their natural habitat, the distinct smell will attract their pollinators, which are flies.

PROPAGATION: This succulent is easy to propagate by stem cuttings. All that is needed is a small piece of stem approximately 1 to 2 inches (2.5 to 5 cm) long. Allow the end to callus over before placing it on a small container of moist potting mix like that which is used for the original plant. The cutting will send out roots in a few weeks.

TOXICITY: There is more than one opinion on whether these are toxic or not, so play it safe and keep away from pets and children.

Orbea variegata (formerly *Stapelia*)

Starfish Flower, Carrion Flower

The stapeliad family of plants are succulents with unusual star-shaped flowers and mottled petals. Though this family is known for their foul-smelling flowers, little aroma emanates from this one. *Stapelia grandiflora* is another story. The aroma from the flowers of this plant brings to mind a dead animal on the roadside on a hot summer day. You may be wondering what makes people want to grow these putrid-smelling carrion plants. One reason is that the flowers are large and unique when compared to other flowers. Or it may be the colors of the flowers that makes them so unusual. Most are a tan, mauve, or burgundy color with unusual markings and texture on the petals. I'm sure the burgundy coloring on the usually tan colored flowers is supposed to resemble blood on a dead animal. These flowers, like all other flowers, are catering to the pollinators they need to pollinate them. In this case, the flesh-colored flowers and smell of rotting meat attract exactly what you think they would—flies.

LIGHT: Carrion flowers are easy to bring into bloom if placed in enough light. A south or west window should suffice for the health of this plant. If they are receiving too much light, the stems will become bleached out and/or will assume a reddish tinge. Plant collectors will actually "sun stress" their succulent plants to bring out these red tendencies as they find it makes for a more attractive plant.

WATERING: Treat these plants like any other succulent, allowing them to dry out quite a bit before watering. Remember, if a plant is in a lower light situation, it will need less water. If it is in bright light or full sun, it will need more frequent waterings. Water consumption will slow down immensely in the winter, so allow your plant to rest, watering it only if their fleshy stems begin to wrinkle.

FLOWERS: If the plant receives enough light, you should have no problem getting it to send out its star-shaped blooms. If it hasn't flowered after one year, move it closer to a light source, be it a window or electric lights. Often these plants are offered as hanging baskets, so the flowers are easier to see. The single flower appears at the end of the stem.

PROPAGATION: These succulents can be propagated by stem cuttings. The carrion flowers naturally have stem segments, and they are easily removed at those points where the stems meet. Let the end callus over for a few days and then place it on a pot of moist potting medium where it will grow roots. Use more than one segment to get a fuller pot of plants sooner.

TOXICITY: *Orbea* plants are not toxic.

Rhipsalis pilocarpa

Hairy-Stemmed Rhipsalis

Rhipsalis are a diverse family of cacti, containing plants with flattened stems resembling a Christmas cactus (*Rhipsalis elliptica*), cylindrical stems (*R. baccifera*), and sometimes hairy stems (*R. pilocarpa*). This plant hails from Brazil, where it naturally grows as an epiphyte, clinging to trees in the jungle. Because *Rhipsalis* are jungle cacti and not desert cacti, they naturally will need more frequent watering, higher humidity, and lower light levels than regular desert cacti. They have small white or yellow flowers in the spring that turn into white, pink, or red berries, offering even more visual interest. Often, the white-berried forms are called mistletoe plants because of the similar-looking berries. It may seem that they are not really cacti because they don't resemble a cactus or have prominent spines. The determining factor that separates succulents from cacti is the areole that the flowers and spines arise from.

Grow these plants in a well-drained cactus and succulent potting medium for best results. Adding orchid mix with plenty of bark pieces will help with drainage and will allow for plenty of oxygen around the roots. Because they naturally grow as epiphytes, they are accustomed to having their roots slightly exposed with plenty of air circulation. They also have shallow root systems like most cacti and epiphytes, so using a short, squat clay pot would be advantageous in this instance.

LIGHT: Since these plants are epiphytic jungle plants, they naturally grow in dappled light that shines through the leaves of the trees they are living on. The environment they naturally grow in gives us hints about what they will need in our homes. Full sun is not needed to get them to bloom like desert cacti. Bright light supplied by an east or west window would be perfect. You could also grow them under lights. Placing them in full sun could result in sunburn on the stems.

WATERING: Don't let these plants completely dry out before watering again. They are used to sudden rain showers that frequently give them water, never really having a chance to dry out, especially with the humidity that surrounds them in the jungle. You may want to stand your plant on a pebble tray to elevate the humidity.

FLOWERS: *Rhipsalis* flowers are quite small, but when the plant is covered in white flowers, it does make a statement. After flowering, small berries appear, which can cling to the plant for some time, adding another season of interest.

PROPAGATION: *Rhipsalis* are easy to propagate from stem cuttings. Make sure they are allowed to callus over before placing them in moist potting medium. Insert the callused cut end into the potting medium or simply lay the cutting on top of the medium, and when it is ready, it will send out roots. Use more than one cutting in each pot to have a fuller pot.

TOXICITY: This plant is toxic to animals. If you grow it as a hanging basket well out of your pet's reach, it should be safe to have in your collection.

Aeschynanthus radicans

Lipstick Plant

When you first see the flowers of the lipstick plant, you will understand how they came by the common name. The bright, colorful, tubular flowers often emerge from a fused calyx, giving them an appearance of a tube of lipstick. These gesneriads are regularly sold as hanging baskets due to their naturally trailing habit. In their native habitat in Southeast Asia, they grow as epiphytes, attached to tree trunks and branches. For this reason, we can safely assume they have shallow root systems that would prefer to grow in a fast-draining potting medium in a shallow container. Though these plants are cousins of the African violet, they would prefer more light than their shade-loving cousins. Since they reside in trees, they are farther up in the jungle canopy and so receive dappled light and may get some peeks of full sun at times.

Most *Aeschynanthus* have red or scarlet flowers, and since the meaning of the genus comes from the Greek words *aischuno* meaning "to be ashamed of" and *anthos* meaning "flower," I guess red flowers were something that caused shame when this plant was named because the color was too flamboyant.

LIGHT: Give this plant bright light to obtain the greatest number of blossoms. An east or west window or hanging it 1 or 2 feet (30 to 61 cm) from a south window would be the best light to promote flowering. Whether it is displayed in a hanging basket or on a plant stand, give the plant a one-quarter turn every time it is watered. If it isn't turned regularly, you may only get flowers on the side that is getting the most light. If your plant starts to develop bleached spots on the leaves, it is receiving too much light and should be moved back from the window. Remember they are naturally growing in the forks of trees and so do not receive full sun. Growing them under lights is also a great way to get this plant to flower more regularly if enough light is not available from the windows.

WATERING: Water this plant thoroughly and then let it dry down a bit, but never allow it to completely dry out. This is imperative, especially when it is flowering, as drying out will result in flowers that may wilt or drop off. Because this plant is from the tropics, elevated humidity is preferred.

FLOWERS: The flowers of *Aeschynanthus* can be red, orange, and even yellow. The flowers arise from an often showy or fused calyx, making it appear as if they are coming out of a lipstick tube. The flowers grow in groups clustered together at the end of the individual vines. They will last a week or more, depending on the temperature, water, and health of the plant.

PROPAGATION: Propagate from stem-tip cuttings inserted in moist potting medium or started in water. If rooted in water, move them to soil when the roots are less than 1 inch (2.5 cm) long. Water roots are different than soil roots, so the sooner they are planted in potting mix, the better they will acclimate. Keep the humidity elevated while they are forming roots by placing them in a plastic bag or cover with a cloche.

TOXICITY: The lipstick plant is not toxic to cats or dogs.

Columnea species

Goldfish Plant, Dancing Dolphins

The common name of dancing dolphins is the perfect description for this plant as the large tubular flowers have the appearance of leaping dolphins. Their flowers have a large lobe at the top of the tube with two small lobes on the side, resembling fins. They are easy to mix up with *Aeschynanthus*, but if you place them next to each other, the difference in the flowers is obvious as *Columnea* don't have the fused calyx that gives the lipstick vine its name and the *Aeschynanthus* doesn't have the large prominent lobe on its flowers. *Columnea* flowers come from the axils of the leaves down the length of the stem instead of being clustered together at the ends of the stems like the *Aeschynanthus*. Like the lipstick plant, this trailing plant will often be found growing in a hanging basket. Goldfish plants are native to Central and South America, where they usually spend their lives as epiphytes in large trees. Their stems will grow roots wherever they touch the trunk of the tree, starting new plants. Make sure your goldfish plant has a well-drained potting medium, and a shallow container is preferable as the roots are usually growing in a small amount of forest debris in trees. Fertilize monthly during its active growing season. It will rest in the winter and won't need fertilizer during that time. These flowering plants may be hard to find at your local garden center, but they are easy to find online from reputable growers.

LIGHT: Goldfish plants prefer a bright light but not direct sun. Because they grow as epiphytes in their native habitat with tree canopies above them, they are more tolerant of lower light situations. To have the best flower production, place it in an unimpeded east or west window. If it isn't blooming in the light you have to offer it, you may choose to grow it under electric lights.

WATERING: Make sure the plant is in a well-drained soil, keeping it moist but not standing in water. Allow it to dry down a couple of inches before watering thoroughly again. When it is flowering, keep it moist or it may decide to drop its flowers and buds may not open completely. Because they naturally grow in the tropics, goldfish plants prefer an elevated humidity. If you have a suitable kitchen or bathroom window, this will be a perfect place for your *Columnea* in the extra humid atmosphere.

FLOWERS: The flowers of *Columnea* are long and tubular, sometimes reaching 3 inches (8 cm) in length. They can be orange, yellow, or red with many combinations of these colors. Their bright colors attract hummingbirds, which pollinate them in nature. Much hybridization has resulted in variegated and burgundy foliage and many combinations of flower colors.

PROPAGATION: Propagate goldfish plants from stem-tip cuttings placed in moist potting medium. They can also be started in water and moved to potting medium after they have sprouted roots. After roots appear, move the plants to a well-drained potting mix.

TOXICITY: *Columnea* plants are not toxic to cats and dogs.

Episcia species

Flame Violet, Peacock Plant

The *Episcia* or flame violet is grown not only for its amazing foliage, but also for its flowers—if grown with enough light. This cousin of the African violet is a beautiful small plant with striking foliage that is often iridescent. It is a stoloniferous plant, like a strawberry plant, sending out small plantlets at the end of its stolons. In its natural habitat, it is a groundcover or may be found growing among rocks and in cracks of walls. Because of this growth habit, they make for a perfect hanging basket so the stolons with their small plantlets can spill over the edge of the pot, showing off its amazing foliage and flowers. *Episcia* can become quite lengthy specimens. These plants, like their violet cousins, have shallow root systems, so growing them in a container such as a bulb bowl or azalea pot is best.

LIGHT: The genus name *Episcia* comes from the Greek work *episkios* meaning "shaded." In their native Central and South American habitat, they would be growing in a shaded area under a canopy of trees. In our homes, the light from an east or west window is best. They can also be grown under electric lights for the best coloration of the foliage and to promote flowering. If your plant has pale leaves, it may be receiving too much light.

WATERING: Keep these plants evenly moist, not allowing them to completely dry out or stand in water. If allowed to dry out, the leaves will turn brown on the edges and the stolons may die back. Humidity is a factor to consider as they come from a naturally humid area. The largely pink-foliaged varieties will need to be grown in an enclosed glass case to keep the humidity high. The other types can be grown outside of glass with perhaps a pebble tray or humidifier used to elevate the humidity. Though it is okay to wet their leaves, do not use cold water because the shock of the water will leave unattractive marks. It is helpful to rinse off the fuzzy leaves with warm water, making sure they dry off out of the sun to prevent burning. If you would rather, a soft paintbrush or makeup brush works well to carefully dust the leaves. Fertilize your plants regularly during the growing season. The small root systems need to provide nutrients to an unproportionate-sized plant for their root size.

FLOWERS: The small flowers of the *Episcia* are mostly found in bright red, orange, or pink colors. These flowers, though not large, are showstoppers against the patterned foliage of the plants.

PROPAGATION: *Episcia* are easy to propagate from the stolons they send out. The small plants can be cut off the ends and inserted in potting medium or can be pinned to a container of potting medium set next to the mother plant. While still attached to the original plant, it will benefit from continuing to get energy from its parent. When it has grown its own roots, it can be cut from the parent. Stem-tip cuttings will also work if stolons aren't present.

TOXICITY: *Episcia* are not toxic to pets.

Kohleria species

Kohleria

These Central and South American relatives of African violets have the same characteristics in that they have fuzzy leaves and stems and prefer the same growing conditions. If the plants become dusty, use a soft paintbrush or makeup brush to dust off the leaves. They can be rinsed off as well, using warm water so the leaves aren't marred. Let them dry fully before putting them back in the sun.

Kohleria are one of the rhizomatous gesneriads, meaning they grow from underground rhizomes. If they experience conditions that are less than stellar, *Kohleria* may go dormant, leading you to believe something went wrong and consequently they died. They are only resting, and you can leave them in the pot and wait for them to wake up, or you can remove the rhizomes from their pots and store them, planting them up again when they show signs of new life. *Kohleria* are fast growers and may become lanky. When that happens, cut the plant back and it will send out new stems. Use the cuttings to propagate new plants.

LIGHT: Give your *Kohleria* bright light. An east or west window would be perfect. If your plant seems to be stretching for the light or does not bloom, it may need to be moved closer to the light source, or it can be grown under lights for almost constant bloom.

WATERING: Like most gesneriads, *Kohleria* would like to be kept moist, but not standing in water, and never be allowed to completely dry out. Drying out may cause the plant to go dormant and die down. Raise the humidity by placing your plant on a pebble tray.

FLOWERS: The flowers of the *Kohleria* are trumpet shaped and comprised of five fused petals. They are called flat-faced flowers because if the flower is placed face down on its petals, it will stand up like a traffic cone. The flowers may be red, pink, purple, or green with many combinations of these colors. They often have polka dots on the inner part of the trumpet-shaped petals. These splashes function as little runway lights for the pollinators—guiding them to the nectar inside the flower.

PROPAGATION: *Kohleria* can be propagated from stem cuttings or from the underground rhizomes the stems emerge from. The wormlike rhizomes can be dug up and split apart into approximately 1-inch (2.5 cm) pieces to start new plants. If you take stem cuttings, make sure there are at least two leaves above the soil line. Make a hole in a moist potting medium with a stick or pencil so that the soft stem is not damaged when inserting it. Damaged stems may allow disease or fungus to enter. Place three or more cuttings into a 4-inch (10 cm) pot and cover with a plastic bag while the cuttings grow roots. Place the cuttings in a bright area but not in direct sun as it may become hot and "bake" the cuttings. They should grow roots in a few weeks and then the plastic can be taken off.

TOXICITY: *Kohleria* are not toxic to dogs or cats.

Nematanthus gregarius

Goldfish Plant, Clog Plant, Guppy Plant

Most of us have childhood memories of owning a pet goldfish. Whether you won one at the local carnival, much to your parent's chagrin, or purchased one at the local pet store, goldfish are a first pet for many children. The endearing goldfish plant will catch your attention as an adult as it is covered with small, orange, pouched flowers resembling that familiar pet. The orange color of the flowers is unique in that there aren't many orange flowers in the houseplant world. Its cascading stems hang from trees as epiphytes in their natural home in Brazilian jungles. That habit makes it perfect for a hanging basket, which is often how it is found for purchase.

The *Nematanthus* name is from the Greek *nema* meaning "thread" and *anthos* meaning "flower." Many *Nematanthus* flowers hang from long pedicels though this particular plant has a short one. You may still find this plant under its old genus name of *Hypocyrta*. Unlike many other gesneriads, there is nothing fuzzy about this plant. Its small, shiny, oval leaves appear to have been covered with an extra glossy lacquer finish. The small, pouched flowers appear on the stems on a short pedicel extending from the axils of the leaves where they meet the stem. The flowers extend along the whole length of the stem for an impressive display. Like many epiphytes, *Nematanthus* have shallow root systems that will grow best in a well-drained potting medium. Add perlite and orchid bark to mimic the growing media of their jungle home.

LEFT: *Nematanthus gregarius* 'Tropicana'

LIGHT: Because these are naturally epiphytes, they are living their best lives in the dappled light of the forest tree canopy. Therefore, they don't need a full sun exposure to bloom well. A west or east window will give the plant all the light it needs to bloom well. It also will bloom well under grow lights.

WATERING: This plant is a little finicky about watering and prefers to be treated like a succulent. Its small leaves are quite thick, so they do have water-storing capabilities. Do not let it completely dry out though, as it will start dropping leaves. Yet, if it is kept too wet, it will also drop leaves and may lose whole stems to rot.

FLOWERS: The flowers have the appearance of the "fish face" that children make when they were young. The plant has gamopetalous flowers, meaning the petals are fused together to form a tube. This is a tube with a big belly. As you may imagine, these tubular, colorful flowers are pollinated by hummingbirds.

PROPAGATION: These plants are easy to propagate from stem cuttings. As you prune your plant to keep it shapely, the cuttings can be used to make more plants. Place the 3- to 4-inch (8 to 10 cm) cuttings into a moist potting medium and keep it moist as the cuttings grow roots.

TOXICITY: The goldfish plant is not toxic to cats or dogs.

Primulina species

Vietnamese Violet

Primulina leaves can be striking on their own, but the delicate flowers that appear to float above the foliage on thin pedicels add to the wow factor. In 2011, these plants were moved from the genus *Chirita*, which was split into seven different genera, and they may be found still labeled as such. *Primulina* are easy to grow and have fuzzy leaves not unlike their cousins, the African violet. They have leaves ranging in shape from round to oblong with plain and patterned green versions. *Primulina* are naturally found growing in limestone areas in China and Vietnam, and they range in size from small forms that grow only 3 inches (8 cm) high to plants that are much larger, some 1 foot (30 cm) or more wide. As these plants grow in rocky limestone areas in nature, they do not have extensive root systems. Keep your plant in tight quarters, being sure not to give it a pot too large for its root system. If the plant cannot use the excess water in the soil in a timely manner, it may lead to root rot. They will be comfortable in the temperatures you are comfortable in, able to withstand temperatures close to freezing, but not liking it.

LIGHT: These are often grown under lights with their African violet cousins but can bloom with less light than the violets. The nice bright morning light of an east window or the light back 1 foot (30 cm) or so from a west window would be best. If they are placed in too bright a light, they may develop burnt patches on their leaves.

WATERING: A well-drained soil is imperative as these are shallow-rooted plants. A pot that is short and squatty is perfect for this plant. *Primulina* are forgiving of drying out but would rather be kept consistently barely moist. Apply fertilizer during the active growing season unless the plant is under lights and then fertilizer can be applied year-round at one-quarter strength.

FLOWERS: The flowers of Vietnamese violets are small and trumpet shaped, usually in pinkish-purple colors, but are also available in yellow and white colors. They rise above the foliage 2 to 3 inches (5 to 8 cm) on thin, dark pedicels. If they are refusing to flower, move them closer to a light source, making sure all other needs are met as well.

PROPAGATION: The propagation of *Primulina* can be done in a couple of different ways. The first involves removing a single leaf and placing it in a moist potting medium, not unlike propagating an African violet. Cover it with plastic or place it in a cloche to keep the humidity elevated. You can also propagate with a leaf off the larger varieties. Cut the leaves into *V*-shaped wedges and place the point of the cutting into the potting medium, making sure to keep the leaf orientation the same as it was on the plant. Small plantlets will appear at the base of the leaf in a few weeks. When they are 1 inch (2.5 cm) or more tall, they can be teased apart and potted separately into small containers.

TOXICITY: *Primulina* are not toxic to dogs or cats.

Streptocarpus section *Saintpaulia* (formerly *Saintpaulia ionantha*)

African Violet

African violets have certainly received bad press in the past, touted as "grandma" plants, boring, and dusty. African violets have come a long way from the time my Grandma Eldred raised them. There were three colors to choose from that included purple, pink, and white. Today through hybridization, the choices of colors, foliage types, and the combinations of the two are almost endless. There are thousands of cultivars now with more varieties coming on the market all the time, and African violet enthusiasts are crazy for them. The best part about them is they are easy to bring into bloom and you can have blooms on your windowsill all year long!

LIGHT: African violet aficionados grow their plants under electric lights, but they are fine on a windowsill and prefer an eastern or morning light exposure. They naturally grow in the mountains of East Africa in dappled shade, so they should not be placed in full sun during the brightest part of the day as the leaves could become sunburned. Under electric light, the leaves will hug the pot if it is too bright, or the leaves may become a dull green and appear bleached out or develop papery brown foliage in spots on the leaves. If this is the case, move your plants farther from the light source. On the other hand, if they are receiving too little light, the leaves will not lie in a flat rosette but will reach for the light. The plant may grow lopsided as it tries to expose itself to the most light possible.

WATERING: There is a controversy concerning the way to water African violets. It is a common belief that you cannot get the leaves of an African violet wet, yet all plants grow outside somewhere in the world. Their life-sustaining water source is rain, that while falling on the ground also wets their leaves. The problem occurs when the water used is cold, which can mar the leaf and shock the root system. I occasionally take my African violets to the sink to rinse off their fuzzy leaves. Every plant enjoys a shower to get the dust off their leaves and debris washed from all the cracks and crevices. If you do water your African violets in this manner, always carefully blot the water out of the center so the plant doesn't develop crown rot. These plants naturally grow in hilly areas on a slight slant, so the water drains from the crown and natural breezes dry the plants after a rain.

Bottom watering is a preferred way to water African violets—never allowing water to get on the leaves. Capillary action draws the water up from the plant's saucer into the growing medium until the root ball is moistened. When the top of the potting medium feels wet, dump any excess water from the saucer so the plant isn't standing in it. When exclusively bottom watered though, soluble salts can build up in the potting medium and negatively affect the plant, so flush water through the growing medium from the top at least once a month to remove the built-up salts and minerals.

(continued)

African violet flowers come in a broad range of colors and variegations.

Wick watering is another way to water African violets as they prefer to be evenly moist. This involves a reservoir of water below the plant that a wick, usually acrylic string, is immersed in. The other end of the string is placed into the bottom of the violet container. The water continually wicks up into the plant, keeping it moist. If this method is used, the potting medium would need to be changed to a fast-draining one. African violets are primarily grown in a peat moss mix, so using the wick water method may keep the growing medium too moist for the plant and root rot may occur.

FLOWERS: Through hybridization, the flowers of the African violet are now quite diverse. There are cultivars with single, double, striped, splotched, and star-shaped flowers. The colors are also diverse, ranging from white to purple, green, pink, and yellow. Their individual flower sizes vary from under 1 inch (2.5 cm) up to 3 inches (8 cm) in diameter and stand up above the foliage in a bouquet that often completely conceals the foliage.

The saucers beneath pots of African violets should not hold standing water as this could lead to root rot.

PROPAGATION: A huge plus for the African violet grower is the ease in which they can be propagated. One leaf is all that is needed. Simply cut the petiole or stem of the leaf to approximately 1 inch (2.5 cm) long, cutting on a slant to create a larger area for root growth. Make a hole before inserting the cutting in a small container of potting medium or an equal mix of vermiculite and perlite so the petiole isn't damaged. Keep it moist, and in six to eight weeks, new plantlets should appear. When those small plantlets are at least 1 inch (2.5 cm) high, transplant them individually into small containers. Many growers offer leaves for sale at lower prices than potted plants, so it is easy to try more varieties.

TOXICITY: African violets are not toxic to pets.

The foliage of *Streptocarpus* is fuzzy, not unlike its cousin the African violet.

Streptocarpus subgenus *Streptocarpella* (formerly *Streptocarpella saxorum*)

Nodding Violet, False African Violet, Cape Violet

The delicate beauty of this plant is magical. The light violet flowers dance above the foliage like tiny, fluttering butterflies. In the past, this plant could be hard to find but now is often offered at garden centers as a shade-loving hanging basket or in small pots destined as a "filler" in a shade container. The natural rounded shape of the plant is perfect in a hanging basket. Unlike many of its gesneriad cousins, it is does not have the low rosulate shape but is more of a low, shrubby, multistemmed plant. Its plump, fuzzy leaves are approximately [illegible] inch (2 cm) long and the flowers are held on thin pedicels 3 to 4 inches (8 to 10 cm) above the foliage. The pedicels seem to disappear, and the flowers appear to be floating above the plant. It is amazing that the thin pedicels can hold up the flowers. The only problem that could be encountered if this plant is grown outside is spots on the leaves caused by cold water from a hose. If possible, water it separately with tepid water from a watering can. It naturally grows in the warm, humid forests of Africa, so it enjoys warm midwestern summers, which are dripping with humidity. If grown exclusively as a houseplant, elevate the humidity with a pebble tray or keep it in a bright bathroom or kitchen where the humidity is naturally higher.

LIGHT: This plant will bloom well in an east or west window or under electric lights. Because this plant is covered with blooms on all sides, make sure to give the plant a one-quarter turn every time you water, exposing the entire plant to the same light. If the plant is never rotated, the blossoms may only occur on the lighted side.

WATERING: Keep your *Streptocarpus* evenly moist. It is forgiving of drying out as the leaves are quite succulent, but it may drop some bottom leaves if allowed to dry out completely. This may result in a leggy-looking plant. If allowed to dry out, the flowers may wilt and drop sooner than normal.

FLOWERS: The flowers resemble small pansies with three large lobes on the bottom and two tiny ones on top. Inside the flower there are dark purple spots, which act as a guide for pollinators to find the nectar. In nature, the pollinator for this flower would most likely be a hummingbird.

PROPAGATION: Because of its shrubby shape, this gesneriad can be propagated from stem cuttings. Trim off a cutting 3 to 4 inches (8 to 10 cm) long, remove the bottom leaves, and insert it into a moist potting medium. Use a skewer or pencil to make a hole first so as not to damage the stems. It will easily root, and by placing two or three cuttings in a 4-inch (10 cm) pot, you will quickly have a full plant. It may become a bit leggy as it ages, so pruning it back will help with that problem.

TOXICITY: This plant is not toxic to cats and dogs.

Streptocarpus subgenus *Streptocarpus*

Cape Primrose

It is exciting to see this previously unavailable gesneriad offered in the annual shade section at garden centers! It is a cousin of the African violet and is a perfect addition to a mixed shade container outside. Previously, it was hardly found and only known as an unusual plant in houseplant collections. Though cape primrose is related to the African violet, their rough-textured, long, strappy leaves in no way resemble the violet's small, velvety leaves. The leaves of *Streptocarpus* are without petioles or stems and arise directly from the soil. The flower stems grow from the bases of the leaves and a plant may be so covered with flowers, the foliage is hard to see.

LIGHT: Give the cape primrose a bright light location but not full sun, which may burn its leaves. An east or west window would be perfect for bringing this plant into bloom. It also does well in a light garden along with African violets.

WATERING: Plant your *Streptocarpus* into a well-drained mix such as you would use for your African violets and most other gesneriads. Purchase a bag of African violet potting mix and add perlite or vermiculite or a bit of both to the commercial mix for a better aerated mix. Water *Streptocarpus* well and then let them dry out almost to the point of wilting before watering again. These plants do not want to have wet feet. Their fuzzy leaves can get dusty and are magnets for pet hair, so use a soft paintbrush or makeup brush to keep them clean. They can be rinsed off under the sink if a tepid water temperature is used to prevent cold water damage on the leaves.

FLOWERS: The flowers arise from the leaves on stems that are 6 to 10 inches (15 to 25 cm) long. Each flower may be up 2 inches (5 cm) across or larger for some of the new hybrids. There are also minute plants with tiny flowers available. The flowers are trumpet shaped, flaring out with a larger bottom lip. They can have slightly ruffled edges to excessive frills that almost hide the center of the flower. The colors can range from white to pink, yellow to purple, and in many combinations of those colors.

PROPAGATION: Remove a leaf and cut it into approximately 2-inch (5 cm) wedge-shaped sections, placing the pointed ends into a container of moist potting medium. In a few weeks, babies will appear at the bottoms of the leaves. Remember, the orientation of the cutting is important. Make sure the pointed side is the side that was closest to the center of the plant. A second way is to lay the leaf on a cutting board and cut away the main midrib so that you have two sides of the long leaf. Place those cut sides lengthwise into the potting medium, and small plants will appear at the bottom of each side vein. Try both ways and see which one works best for you. To create a humid environment, enclose them in plastic or use a small plastic storage container or even a takeout box with a see-through lid. When the plants are at least 1 inch (2.5 cm) tall, move them to their own containers.

TOXICITY: *Streptocarpus* is not toxic to pets.

Gastrochilus retrocallus (formerly *Haraella odorata* and *Haraella retrocalla*)

Fragrant Mini Orchid

The tiny, less than 1 inch (2.5 cm)-wide flowers of this orchid are the most endearing small blooms. They resemble yellow and burgundy pansies complete with "faces." The plant itself rarely exceeds 2 inches (5 cm), but as it matures, it will have more than one set of leaves and so will appear larger. The tiny curving leaves are barely up to 2 inches (5 cm) long and ¼ inch (6 mm) wide, giving an idea of how small the plant truly is. You may wonder why bother to grow something so small. Who doesn't like miniature plants? They don't take up a lot of room and they are adorable. If they are placed in the right environment, these could be in flower more than once a year. This tiny orchid is most often offered as a mounted plant on a cork or tree fern slab. This mimics the way they would be growing as epiphytes in their native Taiwanese (China) home. They may also be found in a small pot planted in a fine-sized orchid bark. Either way is an acceptable way to grow the plant. If you don't have a place to hang an orchid, a potted one may be better for your situation. Just carefully remove the plant from its cork mount and pot it up.

LIGHT: Give your plant medium light to stimulate flowering. An east or west window would be best. If you find your orchid isn't sending out flowers, move it closer to a light source. This orchid would do well under lights as well, especially in an open glass container for elevated humidity.

WATERING: If this tiny orchid is mounted on cork or a fern slab, soak it once or twice a week, depending on the light and temperature conditions. Mine hangs over my kitchen sink and so receives plenty of humidity. It gets soaked at least once a week with other mounted orchids and my *Tillandsia*.

FLOWERS: The tiny yellow flowers of this orchid are pansylike with burgundy markings. They are borne on a short stem with two to three flowers emerging from each stem opening, one at a time. This makes for a longer season of bloom. The large area of the flower is called the lip and it has been determined that the burgundy lip resembles the shell of a beetle, leading to the conclusion that the flower is pollinated by a beetle trying to copulate with the fake beetle. The flowers are supposed to have a citrus fragrance. I've not noticed this, but that doesn't mean a more discerning nose might detect a light scent.

PROPAGATION: This monopodial orchid keeps adding leaves upon leaves and can eventually become up to 3 inches (8 cm) long as it matures. It will send out multiple stems as it ages, and stem cuttings can be taken. Mount it up on its own if roots are present on the stems.

TOXICITY: No toxicity information was found on this orchid, but as it is usually mounted and hanging, it most likely won't be a problem.

Gomesa echinata (formerly *Baptistonia echinata* and *Oncidium brunleesianum*)

Bee Orchid

This small orchid is perfect for a new orchid parent or the person who has limited space for plants. And the cuteness factor is over the top. Who doesn't love flowers that look like bees in flight? No need to fear these adorable bees though, as there are no stingers. This tiny orchid is native to Brazil and was named for the Brazilian ethnologist Baptista Caetano d'Almeida Nogueira. These small epiphytic orchids thrive when they are mounted on cork or tree fern slabs, and most likely this is how you will find this orchid offered for sale. Growing in this manner, with its roots exposed (to an extent), is as close to its natural habitat as can be achieved in our homes. The watering upkeep is a bit trickier as its roots are exposed with only a small layer of sphagnum moss supplying moisture. They may need to be soaked a few times a week to keep their roots hydrated. My *Gomesa* hangs over my kitchen sink, and it thrives with the added humidity my kitchen affords. It blooms reliably for me each spring and sends out more flowers every year as it expands in size.

LIGHT: *Gomesa* need a bright light to flower. A location that is west, east, or back 1 foot (30 cm) or more from a south window would be sufficient. It does not want direct southern sun as this may burn the leaves. If it doesn't flower in the spring, move it closer to the light source. If you observe brown spots on the leaves, it may be too close to the light. This orchid can successfully be grown under electric lights as well and would do well in a glass case to keep the humidity elevated.

WATERING: If your plant is mounted on cork or some other material, it will need to be soaked once or twice a week or perhaps even more. Place it in the sink or a basin of water until the moss and mount are completely saturated. I soak mine for an hour or so and then allow them to drain before hanging them back in their place. This small orchid does have pseudobulbs from which the leaves emerge. These store water, so I find that it is more forgiving of drying out than other orchids. Elevated humidity is preferred, so a bathroom or kitchen window can be utilized provided they have the correct light levels.

FLOWERS: The flowers are grouped together on an arching inflorescence that emerges from the base of the pseudobulbs. Each flower is yellow with maroon markings, and small bracts emerge from each side giving the illusion of "wings" on the bees. The inflorescence can be up to 2 feet (61 cm) long in nature but will reach 1 foot (30 cm) or so in our homes.

PROPAGATION: Separate the pseudobulbs and mount them individually.

TOXICITY: Toxicity for this particular orchid wasn't found, but as it is usually mounted, most likely a pet wouldn't be able to reach it.

The iridescent foliage of the jewel orchid is the perfect backdrop for the flowers.

Ludisia discolor

Jewel Orchid

These are a group of terrestrial (ground dwelling) orchids that have some of the most beautiful foliage in the plant kingdom along with delicate, lovely flowers. Look for the genus *Macodes* for attractive foliage as well. Jewel orchids will need a bit more humidity than other types of orchids and so will work well in a terrarium environment. The first thing you will notice when you encounter a jewel orchid is the glistening foliage. Did someone sprinkle them with fairy dust? When the light hits them, one would think so. The best part is how easy they are to grow and to bring into flower in our homes.

As the jewel orchids are terrestrial orchids, they are grown in regular houseplant potting medium instead of the coarse orchid mix used for epiphytic orchids. Jewel orchids have succulent stems and thick leaves, allowing for more time between waterings, and they are forgiving of drying out. That doesn't mean there aren't consequences for getting too dry. The bottom leaves will turn red, and then brown, and finally cling to the stems if they are extra dry. If this happens often, you may find yourself with a long, naked stem with leaves only on the ends, so keep it moist. Since they are easy to propagate, you can clip those stems off and wait for them to sprout from the cut area to make the plant full again.

LIGHT: Jewel orchids do not need full sun to bloom but instead love a nice bright light such as an east or west window offers. If placed in too much sun, their leaves will burn or fade in color. If that occurs, place it in less light. These plants do well under lights, but when they begin to bloom, move them from under the lights as their flowers rise 12 to 15 inches (30 to 38 cm) above the foliage and may hit the lights.

WATERING: These plants like to be kept moist and thrive with added humidity. Placing it on a pebble tray would be appreciated by your plant. Don't leave them standing in water as the stems are prone to rot if they are kept too wet.

FLOWERS: The delicate white flowers appear from inflorescences at the tip of the plant stem and rise approximately 1 foot (30 cm) above the plant. Many find them a bit anticlimactic, but they appear in the winter, so I find them refreshing and beautiful.

PROPAGATION: Jewel orchids are easily propagated from stem cuttings. As the plant grows, it begins to cascade over the side of the container. The stems are a bit brittle, and if one is knocked off, it is a perfect time to propagate. They can be placed in water, which is kind of fun to watch as the roots are thick with feathery growths. They can also be propagated in soil. Because these plants spread and cascade over, the middle can become bare, and new cuttings can be started right in the same pot. Just add stems to the bare middle of the plant.

TOXICITY: Jewel orchids are not toxic to pets.

Paphiopedilum species

Slipper Orchid, Lady Slipper

The *Paphiopedilum* or slipper orchid is an orchid that should be more readily available because of its fabulous flowers and ease of growth. Its flowers do resemble a slipper, but I like to think it's a ballet slipper rather than a house slipper, which might first come to mind. *Paphiopedilum* orchids are terrestrial orchids, and because they grow on the forest floor, they are more conducive to low light situations as they are only receiving dappled light naturally. Typically, if the plant you purchase has solid green leaves, it will need to be grown in cooler conditions. If you prefer to keep your heater set lower in cold weather, this is the plant for you. Plants with mottled leaves on the other hand like warmer conditions, so keep that in mind when purchasing your *Paphiopedilum*. As with many orchids, the medium they are growing in tends to break down quickly, so even if the plant doesn't need to be up-potted, the medium should be replaced every other year. This ensures the roots are getting the air circulation and oxygen they need to stay healthy.

LIGHT: Slipper orchids can be grown in an east window or back 1 or 2 feet (30 or 61 cm) from a west window. If you only have a north window, you may find they bloom if it is an unobstructed window. They can also successfully be grown under lights, but when the flower bud begins to arise out of the foliage, move it from the light stand so that the flower doesn't hit the light fixture and become distorted. If your orchid refuses to bloom, move it closer to a light source. If the leaves are turning a reddish color or are pale in color, it may need to be moved further from the light.

WATERING: The slipper orchid is frequently offered for sale in sphagnum moss or orchid bark. How often you water will be determined by which mix it is in as well as the amount of light it receives. Temperature will play a role too. They appreciate the temperatures we are comfortable with in our homes, with daytime temperatures near 70°F (21°C) and nighttime temperatures approximately 10°F (5°C) cooler. Some orchids have what are called pseudobulbs (or swollen stems), which store water at the base of their leaves. *Paphiopedilum* do not have those and so will need to be watered more often. Do not allow them to stand in water as that may rot the roots and its leaves will fall off, but don't allow them to dry out either. Keep water out of the leaf axils as that will cause crown rot.

FLOWERS: The slipperlike flowers are held up on a stem from 8 to 18 inches (20 to 45 cm) long and can last for months. Of course, with any flower, cooler temperatures prolong the flowering time. Many of the slipper orchids have glossy flowers that appear to be plastic.

PROPAGATION: As your plant matures, it will send out small clones next to the parent plant. Those small plants can be separated from the parent or left to become a fuller pot.

TOXICITY: These orchids may be toxic to pets, so be cognizant of their placement.

Phalaenopsis

Moth Orchid

The endearing nature of this orchid is the way the flowers float above the plant and look like a group of delicate butterflies or moths in flight. *Phalaenopsis* is derived from a Greek word *phalaina* meaning "moth" and *opsis* meaning "the appearance of." This orchid originated in the tropical regions of the Old World. Younger people may not have been around when the prices of moth orchids were so exorbitant; they were only available to wealthy individuals. Now because of propagation by tissue culture, moth orchids are available at almost every garden center and grocery or big box store at much more reasonable prices. The mothlike flowers come in an array of colors, including ones that look unnatural. If you find the bright blue or extra vivid orange or green colors of moth orchids, they may be dyed. When they bloom again, flowers will be either white or the color they were before they were dyed. The process involves injecting dye into the stem of the flower to allow the color to travel through the stem into the petals of the orchid flower.

Phalaenopsis orchids are great houseplants because they are readily available, affordable, and easy to rebloom.

LIGHT: Moth orchids don't need a bright light to bloom. An east or west window should bring them back into bloom, or growing them under electric lights also works. Many feel that after they are done blooming, these plants are disposable and throw them out. I've been given many orchids in the past from friends who knew I would take them, so they didn't have to throw them out. If composting them works for you, don't feel bad. Often buying an orchid plant gives you flowers longer than a disposable cut flower arrangement. Yet, with good care and the right light, you can get them to bloom again.

(continued)

When purchasing a *Phalaenopsis*, choose one that has buds on it to extend the bloom time.

WATERING: You will find that when you buy an orchid, it is usually in a decorative pot cover (cachepot) without a drainage hole. The orchid itself is growing in an opaque plastic pot that has drainage. To water the orchid, take it out of the decorative pot, water it in the sink, let it drain, and return it to the cachepot. Watering them while they are in their decorative pot can cause problems. Because there is no drainage, your plant might end up standing in water, which will eventually kill the plant. When watering the orchid, run water through the potting medium along with wetting any roots that may be outside the container. Those roots need to be moistened as well as the ones in the potting medium. A well-watered moth orchid root is a greenish color, and when they are dry, they are a silver-gray color. The see-through pot allows you to see the roots and determine if they are dry.

Though the entire plant can get wet, if water is left standing in the center of the leaves, this could cause the leaves to rot and the plant to fall apart. Trust me. If water is in the center of the leaves after watering, use a paper towel to carefully soak up the excess water. In nature, these plants would be growing as epiphytes on trees or other objects and most likely wouldn't be growing straight up and down, so excess water would drain out. They also have the natural wind outside to dry them off.

Phalaenopsis orchids last many months as long as the plant isn't allowed to dry out.

Phalaenopsis orchids need to be repotted in fresh orchid mix at least every second year. The bark or moss they are potted in begins to break down and then the roots aren't getting the air circulation they need. While repotting them, cut off any brown, mushy roots, leaving the firm green ones.

There are lots of mini orchids that take up less room but still pack a punch of color.

FLOWERS: The flowers of the moth orchid are produced on a long flower spike that emerges from among the leaves of the orchid. The flowers can emerge opposite each other on the spike (called a raceme) or branch out from the main spike (the panicle). The flowers open from the bottom up with the last flower emerging at the end of the spike. These flowers can last for months if the plant is well watered and taken care of. When they are finished flowering, after the spike is brown and shriveled, cut it off at the base of the spike. Then the plant will put its energy back into itself so that it will have a better flower display the next year. When purchasing a plant, buy the one with the most buds instead of the plant with every flower fully open so you will have a longer display.

PROPAGATION: Propagation is now mostly done in laboratories but can still be done at home if a keiki (baby plant) appears. Sometimes they pop up next to the mother plant or may appear on a flower spike. Once the keiki has roots that are approximately 3 inches (8 cm) long, you can cut it off and pot it up in its own container.

TOXICITY: Moth orchids are not toxic to pets.

Cyclamen persicum

Florist Cyclamen

The cyclamen plant is native to the Mediterranean area, so its bloom time coincides with the late fall and winter holidays in the Northern Hemisphere. Most are purchased between Thanksgiving and Valentine's Day. The plant will keep growing through the spring but then begins to go dormant. They usually rest in the summer months after the foliage dies down in the late spring, which has led many to believe that the plant has died and it is discarded. *Cyclamen* grow from an underground tuber, where it stores energy to withstand the drought and heat of its native land during the summer. When the rains return and the temperature cools down, it begins to grow again. We can mimic this dormancy by storing the pot away from the light for the summer, making sure the tuber doesn't completely dry out, and bringing it back out in the fall when it starts to resume growth.

The hot-colored flowers range from red to pink to white, sometimes with fringed edges and a slightly sweet fragrance. The foliage alone is worth buying the plant for as it is heart shaped and attractively marbled. If your plant is kept too warm, it will begin to display yellowing leaves and die down early.

LIGHT: *Cyclamen* need a bright light to flower but never direct sun as this may burn the foliage. This plant naturally grows as a groundcover in its native Mediterranean area. An east or west window would be preferable.

WATERING: *Cyclamen* like to be kept moist, but never standing in water, which will rot the roots and possibly the tuber the leaves emerge from. It is best to carefully water around the edges of the pot, avoiding watering in the middle of the plant where water can sit and rot the middle of the tuber. If the tuber is kept too wet, the flowers, and possibly the leaves, will wilt and could ultimately cause the death of the plant. Bottom watering is another option, setting the pot in a bowl of water and letting it soak up the water it needs. When the top of the potting medium feels damp, remove the plant from the water. Make sure to fertilize it lightly during its growing season to help replenish the tuber.

FLOWERS: The flowers of *Cyclamen* have five reflexed or curved back petals that are joined together at the bottom. The flowers are often compared to butterflies, badminton shuttlecocks, or birdies and form a dense bouquet that stands up a few inches above the foliage. As the individual flowers die, remove them by sliding two fingers to the bottom of the flower stem, give them a quick twist and pull. They should come away easily from the plant.

PROPAGATION: Propagation of *Cyclamen* is best from seed, but the process can take over a year. The tuber can also be cut up and planted individually, but this is not recommended for the typical plant parent. It would be best to buy another plant rather than trying to start more from an existing plant.

TOXICITY: *Cyclamen* is toxic to cats and dogs.

Euphorbia pulcherrima

Poinsettia, Christmas Star, Mexican Flame Leaf

Nothing is more iconic at Christmas than the poinsettia plant. Its common name of Christmas star is fitting as it shines like a star during the holiday season. The species' Latin name *pulcherrima* translates to "most beautiful," setting it apart from other *Euphorbias*.

The poinsettia is a plant that is either loved or hated, but everyone can agree it does represent the Christmas season. Sold for approximately six weeks around the holiday season, remarkably it is the top-selling plant in the world. That is a lot of plants that need to be ready and looking pristine in a short time span. The poinsettia is a short-day plant needing long nights to trigger the colorful bracts and flowering. This means it needs fourteen to sixteen hours of darkness every day for eight to ten weeks for them to bloom. Growers must have the timing and conditions perfect as even the smallest amount of light during the middle of the dark period would be devastating and the plants may not color up.

We have Joel Roberts Poinsett (1779–1851) to thank for bringing these plants to the United States from Mexico in the 1820s. Mr. Poinsett was an amateur botanist and the first U.S. Minister to Mexico, and when he saw the poinsettia, he sent cuttings back to his home in South Carolina. Fast forward many years to the 1900s, and the spotlight is on the Ecke family who took the naturally straggly growing poinsettia and turned it into the lovely, compact, multibranched plant that we now enjoy in our homes. Many poinsettias around the world begin their life in California on the Ecke ranch and what once only came in red are now offered in white, pink, burgundy, and speckled and splashed forms. Don't worry if you don't find a color that suits your indoor décor because they can be spray-painted any color your heart desires. As Joel Poinsett passed away on December 12, in 2002, Poinsettia Day was created honoring Joel and the Ecke family. In Mexico, poinsettias are known as La Flor de la Nochebuena, or Flower of the Holy Night.

Poinsettias are offered in a multitude of stores during the holiday season. Though they may all look the same to some, there are subtle differences, and you will want to pick the one that will last many weeks in your home. Choose one without any yellowing leaves at the bottom, which indicates they may have dried out, have been inconsistently watered, or are standing in water. The latter is usually the case as most are offered in foil pot covers that have no drainage. Check the inside of the pot cover to see if the plant is standing in water before putting it in your cart.

(continued)

Check the pot cover before buying your poinsettia to make sure it isn't full of water.

Poinsettias have been hybridized and now are available in many beautiful colors.

If you see a poinsettia still sleeved, be wary as poinsettias should be removed from their shipping boxes or sleeves immediately after arrival at their destination. Ethylene gas can build up in the enclosed spaces and cause epinasty or wilting of the poinsettia from which it cannot recover. When you bring your poinsettia home, it should be sleeved in a paper sleeve (or large plastic bag at the very least). Poinsettias do not like temperatures below 50°F (10°C) and can quickly become damaged.

LIGHT: When you bring a poinsettia into your home, find a place with bright light and warm temperatures. Do not place your plant near a fireplace or heat vent though as the heat will age them more quickly and may even damage them if the heat is excessive.

WATERING: As stated above, most poinsettias arrive in decorative pot covers. When watering them in your home, remove the plant from the pot cover, take it to the sink or place it in a saucer to catch the water, and moisten the plant thoroughly until water runs out the drainage holes. Let it drain, then return it to the pot cover or you may choose to plant your poinsettia directly into a decorative pot (as long as it has a drainage hole). They are susceptible to spider mites in dry winter home environments, so keep the humidity elevated around them.

Poinsettias have many different colors of bracts while the true flowers are the yellow structures in the middle.

Even if your poinsettia never turns color again, it is a beautiful houseplant with its red petioles and deep green leaves.

FLOWERS: Though it seems the colorful poinsettia has enormous flower petals, these aren't flowers but are bracts or modified leaves. The true flowers are the tiny yellow structures in the center of the colorful bracts and are collectively called the cyathia.

When purchasing a poinsettia, make sure those flowers in the middle of the bracts are still a light yellow color. These plants are fresher, younger, and longer-lasting poinsettias. If the pollen grains are showing yellow on top of red filaments, the flower is nearing the end of its life and the bracts will soon begin to fall. Bringing a poinsettia into flower the next year is possible but does involve the darkness period discussed earlier. Poinsettias do make a beautiful houseplant, though.

PROPAGATION: Poinsettias can be propagated from stem cuttings and in fact make great cut flowers. Leave them in the vase and they will begin to root, and later you can pot them up.

TOXICITY: This plant is not toxic to pets, but the latex sap may be irritating to some people.

Hippeastrum

Amaryllis, Knight's Star Lily

During the holiday season, the amaryllis has no rival for its beauty. The enormous flowers come in colors ranging from white, to red with striped and variegated versions as well. When purchasing this plant during the holiday season, try to find loose bulbs to buy as this is the preferable way to buy an amaryllis. The potting medium in the boxed versions isn't conducive to good root growth. Invest in the largest bulb available as it may send up two to three stems with as many as six flowers on each stem. Buying a smaller bulb may not yield a flower at all if it isn't mature enough. The larger bulb will be worth it in the long run.

After purchasing the bulb, soak the roots of the amaryllis in a glass of water for an hour or two before planting. Don't let the bulb touch the water, just the roots.

These bulbs have been stored for some time before they get to you, so the roots are often wrinkled and desiccated. The roots may be quite lengthy, so try to find a tall, narrow pot to house your bulb. Plant in a well-drained potting medium in a pot that is only 1 inch (2.5 cm) wider than the bulb on either side.

The top one-third to one-half of the bulb should be showing above the potting medium. Water it lightly, place it in a bright, warm spot, and wait for the bulb to come to life.

As your bulb begins to grow, it will often send out leaves at the same time as the large inflorescence. Or it may send out the flower stalk first and the leaves will appear later. Either way is fine. If leaves are the only life to appear and it doesn't flower, that simply means your bulb wasn't mature enough to flower this year. Let the leaves grow and nourish the bulb, and next year it may be big enough to bloom.

(continued)

Soak the roots of your amaryllis in water before planting.

Use a well-drained potting mix for your amaryllis. Grow it in a heavy container no more than 1 inch (2.5 cm) wider than the bulb on either side.

Notice the bulb is planted almost one-half of the way out of the potting medium and the pot isn't much larger than the bulb.

LIGHT: These tropical South American bulbs prefer a bright light to do their best. If you have purchased a large amaryllis bulb, all the energy the bulb needs to flower and send out leaves is already present in the bulb. Therefore, the light isn't that important while it is flowering. You could place it anywhere you want color, and when the leaves appear, move it to the light it needs to grow well. Giving it bright light is important to keep the leaves more compact and sturdier. If it isn't getting enough light, the leaves will stretch and flop. Move the plant outside when the danger of frost is past and grow it in a semi-sunny spot. Full sun might be too much for it, so keep it where it won't receive afternoon sun. Fertilize your plant all summer while it is actively growing to help rejuvenate the bulb.

WATERING: After you plant your bulb, water it in to settle the potting medium around the roots. Place it in a bright, warm spot and don't water it again until you begin to see growth. Until green growth appears, the plant won't have any photosynthetic capabilities, so its water use will be minimal. As it begins to grow, keep your plant moist, being careful not to get water into the neck of the bulb.

If your flower stalk breaks, place it in a vase and it will last for a few weeks.

FLOWERS: The flower stalk usually appears first and will immediately find the light and start growing toward it, so you will need to turn the plant often to keep the stem growing straight. When the flowers begin to open, make sure the plant is not allowed to dry out as that will shorten the life of the flowers. Sometimes the weight of the flowers may cause the inflorescence to fall over. If this happens, either stake the stem or cut the flower stalk and place it in a vase of water. Amaryllis makes a wonderful long-lasting cut flower as well.

PROPAGATION: As your bulb ages, it will begin to grow small bulblets at the base. These can be separated from the mother bulb and put in small pots of their own. If you choose to leave them on the original bulb, eventually you will have a large pot full of flowering bulbs. Keep moving it into a bigger pot that is only 1 inch (2.5 cm) or so wider on either side of the bulbs.

AFTER CARE: At the end of summer, allow your plant to dry out and let the leaves die down. You can remove the bulb from the pot or leave it as is. Cut the yellowing leaves off and allow the bulb to rest for at least two months. Then six to eight weeks before you want it to flower again, start watering the bulb and wait for it to send out its next display of flowers.

TOXICITY: The amaryllis bulb and plant are toxic to pets.

Kalanchoe blossfeldiana

Flaming Katy, Florist Kalanchoe

A blooming *Kalanchoe* is a plant that makes people happy. The flowers come in a range of colors, including red, pink, orange, yellow, and white. These aren't washed out, pastel colors; they are bright, vibrant, beautiful colors and the reason for the common name, flaming Katy. This plant is available for sale year-round but is most often found during the holiday seasons, including Christmas and Easter when they are often purchased as hostess gifts. They are short-day plants like poinsettias and will start to bloom after a few weeks of fourteen-hour nights. They can be grown all year for sales as greenhouses can simulate the short days and long nights the plants need to be brought into bloom.

Some gift plants are enjoyed while they are blooming and then tossed in the compost pile. I hope after reading this, you decide to keep your *Kalanchoe* and try to bring it back into bloom. Many people don't realize that kalanchoe is a succulent. Under all those flowers are dark green, fleshy, scallop-edged leaves. Even if the plant never flowers again, it is a beautiful, succulent, foliage plant. There are variegated foliage types as well.

LIGHT: For maximum blooms, this plant would prefer to be in full sun. An unimpeded south or west window would be best. It can also be grown under electric lights.

WATERING: This Madagascar native is a succulent and would like to be thoroughly watered and then allowed to almost completely dry out. If the leaves appear wrinkled, it has been allowed to become too dry. The number one killer of this plant is root rot from allowing the potting medium to remain wet for too long. Water it thoroughly, but empty any water left in the saucer after thirty minutes, making sure to use a fast-draining potting medium. Using a clay pot to house your *Kalanchoe* is a good idea as it allows water to escape through the walls of the container as well, helping prevent root rot from too much moisture.

FLOWERS: When purchased, this plant will be covered in small, four-petaled flowers, with most of the foliage hidden from sight. Usually, the plant cannot be brought back into flowering with such a large head of flowers, but even with less flowers, it is well worth growing. If it becomes lanky and tall, cut it back to keep a more compact, bushy form. As the fall days become shorter and the nights longer, this change should stimulate the plant to bloom, but you could help ensure this by covering the plant with a box fourteen hours a day for a few weeks to prompt the flowering process.

PROPAGATION: This plant can be grown from seed or from stem cuttings. If using a stem cutting, allow the end to callus over first before planting it into a well-drained potting medium. In a few weeks, it should have roots and begin to grow new stems.

TOXICITY: *Kalanchoe* are toxic to pets.

Oxalis triangularis

False Shamrock, Good Luck Plant, Wood Sorrel

The popularity of *Oxalis* may harken back to childhood days of seeking out four-leaf clovers in the grass. My grandma was a master at finding them, and we have discovered countless pressed four-leaf clover leaves in her books over the years. Though these from the lawn are in no way related to the *Oxalis* we have as houseplants, the resemblance is there. You will find *Oxalis* or shamrocks for sale in March for St. Patrick's Day, but you also may find them in the spring in the annual section at the garden center. The endearing foliage resembles butterflies as they seem to float on their thin stems. The cool thing about this plant is that its leaves are an example of photonasty. You may wonder what that means—it doesn't sound nice. It simply means the plant responds (in this case very dramatically) to light changes. When the light levels dim, the plant folds up its leaves and rests when it has no light. It goes to "sleep" at night like we do. When the light returns, it wakes up and its leaves return to their former state.

If the light or moisture levels aren't sufficient or it is too warm, this plant will wilt, its leaves will turn brown, and it will go into a dormant state. It may do this naturally after flowering too. Often people mistake this for the death of their plant and throw it out. Yet, if they had been patient, the plant would have sent up new leaves and become a beautiful, full plant again. This dormancy may last from weeks to a month (or two), and when it happens, the plant needs to be stored in a cool, dry place.

LIGHT: Your *Oxalis* needs bright light to full sun exposure to stay compact and not become leggy. If the leaves are stretching toward the light excessively, move it closer to the light source. If it becomes leggy, the leaves will collapse over the side of the pot as the stems can't hold up the leaf. A west or south window would be the best exposure, or it can be cultivated under grow lights. If it develops brown patches on its leaves, it may be experiencing too much sun and will need to be moved back from the light.

WATERING: Plant your *Oxalis* tubers in a well-drained potting mix. If the potting medium is too heavy (and holds too much water), the tubers may rot. If the plant is allowed to completely dry out, it will most likely send the plant into dormancy. Only allow the top 1 or 2 inches (2.5 to 5 cm) of the potting medium to dry out before watering again.

FLOWERS: Depending on the species of *Oxalis* you choose to grow, the flowers can be pink, white, and even yellow. The delicate, five-petaled, bell-shaped flowers rise above the foliage on thin pedicels.

PROPAGATION: These plants can be propagated by splitting the plant. They grow from underground tubers, and they can be separated and potted up separately.

TOXICITY: While *Oxalis* leaves are edible, eating too many may be upsetting to your stomach. A sprinkling of leaves on your salad would be sufficient. The leaves contain oxalic acid, so it is best to keep it out of the reach of pets.

Schlumbergera species

Holiday Cactus, Christmas Cactus, Thanksgiving Cactus

Holiday cacti are a quintessential part of the holiday season. Not unlike the poinsettia, you can find these plants in almost every store during November and December. Though the plants found today are exclusively hybrids of the Thanksgiving cactus (*Schlumbergera truncata*), the Christmas cactus (*Schlumbergera* × *buckleyi*) can occasionally be found to purchase in smaller, older greenhouses. They aren't the same, so let's look at the differences. Mainly the difference is in the shape of the cladodes or flattened leaflike stems that do the photosynthesis work that leaves perform for other plants. The Thanksgiving cactus, also called the claw cactus, has small points on the segments, whereas the Christmas cactus has rounded edges. Often, large, decades-old Christmas cacti, with their bright pink flowers, have been passed down from generation to generation.

Though it has been argued that these plants aren't true cacti, the presence of areoles sets them apart as cacti. Because they are jungle cacti hailing from the Brazilian rainforest, they prefer high humidity and consistent moisture in a well-drained potting mixture. They grow as epiphytes on other plants, using them for support rather than taking any nourishment from them, thus they are not parasites. They can also be found growing as lithophytes in the crevices of rocks.

The genus *Schlumbergera* has gone through many name changes from *Epiphyllum*, to *Rhipsalidopsis*, to *Zygocactus*, to the name it has now in honor of Frédéric Schlumberger.

The other popular holiday or jungle cacti is the Easter cactus (*Schlumbergera gaertneri*, formerly known as *Rhipsalidopsis* or *Hatiora*), which blooms in the spring. Its flowers are star shaped, and it has small bristles protruding from the areoles. It is available in the original red, orange, white, and pink colors. As it comes out of our winter, the longer and warmer days help trigger flowering.

(continued)

The true Christmas cactus has cladodes with rounded edges and fuchsia pink flowers.

Stephanotis floribunda

Madagascar Jasmine, Bridal Flower

You may have encountered *Stephanotis* without realizing you had been introduced. The meeting was most likely at a wedding in a bouquet or boutonniere. More likely, your nose has encountered the heavenly fragrance that smells like a jasmine, thus the common name. This tropical vine is commercially grown for the clusters of white, waxy flowers that are used in the cut flower industry. These vines, though commonly called jasmine, are not related at all but are in the milkweed family—making them closely related to the *Hoya*. If you find a *Stephanotis* to purchase, it will commonly be offered for sale on a hoop trellis as it is a vining plant. In its natural habitat in the tropical areas of Madagascar, it is a rambling vine that can grow over 20 feet (6 m) long. This vine would prefer having elevated humidity with plenty of light and water. When it is dormant in the winter, cut back on the water.

LIGHT: Give this plant bright but not direct light. An east or west exposure would be appreciated. If you place it in a south window, move it back 1 to 2 feet (30 to 61 cm) so its leaves are not receiving direct sun exposure. This may cause the leaves to develop brown areas of burnt tissue. If your vine doesn't bloom, move it closer to the light source or try growing it under electric lights if you can't find an appropriate place for it.

WATERING: While this plant is actively growing, it would like plenty of water as long as it is residing in a well-drained potting mix. Make sure that it isn't allowed to dry out while flowers are present as this may cause the flowers to drop off the plant prematurely. Fertilize monthly while it is actively growing, stopping when the plant is dormant in the late fall through winter.

FLOWERS: The flowers are pure white and have a shiny, waxy appearance with five petals that flare out like a trumpet. Flowering takes place in the spring and summer months when there is enough light and water. *Stephanotis* fragrance is strong, so this may not be the plant for your home if someone in the family has a sensitive nose. Keep your *Stephanotis* a bit snug in its pot to promote flowering.

PROPAGATION: Propagate by stem-tip cuttings in the spring and summer. Place the cuttings in a moist potting medium or root them in water. Place three to five cuttings in a pot to make for a fuller plant. If you want to train the vine around a trellis, place the trellis in the pot when the plant is young so it can be trained from the start.

TOXICITY: *Stephanotis* is not toxic to cats or dogs. It does bleed a white substance when cut open, so be aware of that when pruning your plant. Wear gloves and make sure it doesn't drip on your furniture.

Anthurium andraeanum

Flamingo Flower, Painter's Palette, Tail Flower

Many *Anthurium* are prized for their amazing foliage, but the flamingo flower is more well-known for its patent-leather-like inflorescences. At one time these plants were only found in red and white, but today, the color palette is much larger and includes green, purple, pink, and striped versions to choose from—with much hybridizing still being done.

LIGHT: This plant needs a bright light to bloom but can become sunburned in too much light. An east or west exposure is best, steering away from a southern light exposure. *Anthurium* often grow as understory plants in their native jungle homes in Colombia and Ecuador. These plants also may grow epiphytically on other plants and so are receiving dappled sun in their natural habitat.

WATERING: Always keep the potting medium of *Anthurium* moist. If allowed to dry out, the plant will develop yellowing and dropping leaves. At the same time, never leave it standing in water as this is a sure killer of your *Anthurium*. A potting medium that is quick draining is best. Because they often grow as epiphytes in their native home, they would be growing in the forks of trees with forest debris as their main "potting medium."

FLOWERS: Though the large showy part of the inflorescence is often thought to be the flower, this is a spathe or modified leaf. The small tail that sticks up from the spathe is the flowering structure of the plant. It is made up of countless small flowers packed together on the small protuberance. The genus name *Anthurium* comes from the Greek words *anthos* meaning "flower" and *oura* meaning "tail," which refers to the tail-like spadix. The inflorescence can last for months on the plant with the right care and may be in continuous bloom throughout the year, providing color in your indoor garden.

PROPAGATION: *Anthurium* can become tall and leggy. When living in the jungle, it would need to compete for the dappled light coming down through the trees. Therefore, it grows almost like a vine, sending out small aerial roots to help collect moisture. If your plant is growing like this, it can be staked or cut, and the cutting can be used to propagate the plant. If your plant isn't too leggy but is showing a bit of "ankle," the plant can be repotted, sinking the leggy stem lower into the medium where the aerial roots will begin to grow. This may involve cutting off a slice of the roots from the bottom of the plant to match the length of the stem or neck you want to cover. If the stem is a couple of inches long, trim the plant back and root the cuttings in either water or potting medium. Don't throw out the root system of the mother plant; you may find it will sprout new stems and the plant can start up again.

TOXICITY: This plant is toxic to pets and humans. The tissue of the plant contains calcium oxalate crystals released when the plant is being nibbled. These could cause irritation in the mouth, and if swallowed, in the intestines.

Aphelandra squarrosa

Zebra Plant

When you first see this plant, you will understand the reason for the zebra moniker. The bracts of this decorative plant are the star of the show, resembling Japanese pagodas. The bracts are stacked on top of one another in a yellow conelike shape. The small, yellow flowers peek out of those layers of bracts. The prominent white stripes on the dark green leaves are very striking when the plant isn't flowering. *Aphelandra* comes from tropical Brazil where it grows in the understory.

It does well in low light homes but does need to have close attention paid to it as this plant has a reputation of being a hard plant to grow. If it is kept evenly moist with no big fluctuations in the moisture and the humidity is elevated, it will do well. Just make sure the plant does not dry out completely or the leaves will drop. They will also drop if the temperature is too cold or the plant is kept too wet, especially in conjunction with cold temperatures. Keep it away from cold windows or drafty areas in the cold months. Because of its dark green colored leaves, we can assume it is more of an understory plant in its native habitat. Dark green leaves have more chloroplasts to collect as much light as they can for photosynthesis—as the dappled light peeks down through the leaves above them. *Aphelandra* can become somewhat straggly plants, especially if they have been allowed to dry out and the bottom leaves fall off. In this case, it is easily pruned back and will grow new foliage. Make sure to elevate the humidity around this plant or it may develop a case of spider mites, especially when the heater turns on for the winter, drying out the air. If you have a humid bathroom with good light, the zebra plant would do well in this situation.

LIGHT: Place this plant in a bright light without full direct sun. Because it has thin leaves, it can become sunburned in too much sun. For it to flower, it does need to get some good light, so an east or west window would work well.

WATERING: Do not let this plant completely dry out as leaf drop is the consequence of neglect. It also does not want to stand in water. Pot it in a well-drained potting medium with plenty of peat to help the potting medium retain moisture better.

FLOWERS: The most obvious part of the flower inflorescence is the colorful yellow bracts (modified leaves) that rise from the tips of the plant. The actual flowers, which are small and yellow, emerge from between these bracts. The bracts act as a beacon to attract pollinators to the small flowers.

PROPAGATION: This plant can be propagated from stem-tip cuttings rooted in water or placed in moist potting medium. After it blooms is a good time to give it a trim and use those trimmings for propagating.

TOXICITY: This plant is nontoxic to cats and dogs.

Begonia species

Begonia

When Michel Bégon (1638–1710) discovered a begonia in present-day Hispaniola, little did he know how popular this group of plants would become. The begonia type that many are familiar with as a houseplant is the beautifully leaved Rex begonia, though outdoor bedding begonias purchased in flats are the best known. Rex begonias are grown for their foliage yet do bloom, but with sparse blooms on short pedicels that often never rise above the foliage. So, let's concentrate on the begonias that are known for their amazing blooms in the house. The best part is that these begonias send their flowers out during the short days of winter when not much else is in bloom. Those delicate sprays of flowers are a wonderful, much needed pick-me-up after the busy holidays.

The indoor begonias include the cane and rhizomatous types. Many of these still have gorgeous leaves, but their blooms are much more pronounced than the Rex varieties. The cane begonias are often called angel wing begonias, and the popular *Begonia maculata* is a good example of one. Cane begonia stems have the appearance of bamboo and so is the reason for the common name cane begonias. They can bloom continuously with the right light and growing conditions. They have large clusters of flowers that dangle from the canes with pink, red, or white petals.

Rhizomatous begonias grow from large creeping stems called rhizomes, from which the foliage arises from the top and the roots grow from the bottom. The rhizomes are visible above the soil line, and older plants may have rhizomes that creep right over the rim of the pot and keep going. Shallow, large, bowl-like pots work great for rhizomatous begonias so they can crawl along potting mix instead of falling over the side. Whichever type of begonia you choose, the foliage and flowers are sure to be showstoppers.

LIGHT: Most begonias are assumed to be shade loving and many are, but they like a bright light, and the cane begonias can take some direct sun even from a southern exposure. An east exposure with soft morning light for the rhizomatous varieties and a west exposure for the cane types will ensure blooms on both types of plants. If you find they aren't producing any flowers, move them closer to the light source. If you don't have adequate light, think about adding supplemental electric lights to boost the light levels and get them to bloom.

WATERING: Being perplexed by the watering needs of begonias is the number one reason for failure with these plants. They have shallow root systems, so plant them in shallow containers such as azalea pots. Some of the cane begonias can get quite large, so they may need deeper containers as they can become quite tall and top heavy. Make sure the soil isn't too heavy (water-holding). Using clay pots is desirable as this type of pots let extra moisture escape through the walls of the container. Water your plant thoroughly until water runs out the drainage hole. Allow them to become dry 1 or 2 inches (2.5 to 5 cm) down before watering again. These begonias have succulent stems and petioles, so they have water-storing qualities.

(continued)

Female begonia flowers have a winged ovary.

This *Begonia* 'Jackie Brookshire' is a striking begonia with peach-colored flowers.

FLOWERS: The flowers of most begonias are monoecious, meaning that the male and female flowers are on the same plant. You will be able to tell the difference as the female flowers develop a winged ovary where the seeds are formed. This will turn brown as it ripens. Most begonias have flowers in colors from white to red with all shades of pink in between.

PROPAGATION: Propagation of begonias depends on the type of plant, and there are numerous ways it can be accomplished. Stem cuttings can be used for the two types of begonias we are discussing. As begonias can become almost woody, cut a piece from the tips where the cane is still relatively soft. Insert it into moist potting medium and the cutting will root within a few weeks to a month. Rhizome cuttings can be taken from the rhizomatous types. Cut the tip of a rhizome, preferably with two or more leaves intact. The cutting needs only to be laid on top of moist potting medium with the bottom half of the rhizome under the soil. A curved piece of wire can be placed loosely over the rhizome to keep it from falling over. The cut end of the rhizome cutting can also be placed down into the potting medium like the upright stem cutting. Single leaf propagation can also be used for both types of begonias. One petiole with the leaf intact can be used placing it in potting medium or in a receptacle of water.

Begonia flowers are usually red, pink, or white like these above.

The leaf can also be cut into wedges, with each wedge containing a larger leaf vein section. Stick the wedges into barely moist sterile medium and cover it with plastic or a cloche to keep the humidity elevated. A clear plastic shoe box or deli container can also be used to start multiple cuttings. Within a few weeks, small plantlets will begin to appear. Remove the covering when the new plants are almost touching the plastic.

TOXICITY: Begonias are toxic to cats and dogs.

Citrus species

Citrus

There are many different citrus varieties to choose from, and selecting one that is known to be easy to grow and bloom inside is the best choice. Some of the easiest to grow inside are the calamondin orange *Citrus* × *microcarpa* (formerly × *Citrofortunella mitis*), the Meyer lemon *Citrus* × *meyeri*, kumquat *Citrus japonica* (formerly *Fortunella* sp.), and the Persian lime *Citrus* × *latifolia* (formerly *Citrus aurantifolia*). Most citrus trees are grafted onto a dwarf rootstock to allow the upper plant to stay smaller. This is done as many citrus can reach heights and widths of over 20 feet (6 m). The aroma of a citrus plant in bloom is unmatched by any other plant. The fragrance is intoxicating, and who doesn't want their late winter home to smell like spring? *Citrus* may flower but not fruit if they are not given a bit of help with the pollination. Shaking the tree or using a small brush and brushing pollen between a few flowers will help. Do this carefully as some citrus have thorns on their stems.

LIGHT: To get the optimal amount of bloom, make sure your citrus is in the most light you can provide it. Growing them directly in front of a south or west window should work. If there aren't any flowers during the winter or spring, move it where it will get more light or add some supplemental light over top of the plant. Citrus plants would appreciate a summer on the patio too.

WATERING: Watering your citrus plants is something that needs to be done thoughtfully. They never want to be overly wet or have "soggy bottoms." It would be best for the plant to be grown in a clay pot, which allows for excess moisture to evaporate out the sides of the pot. It is also easier to see if the soil is moist at the bottom as the clay will be a darker color there. If your plant is in a large pot, use a dowel to check the moisture at the bottom of the pot. Push the dowel in, leave it there for a short time, pull it out and if the end is wet, don't water your plant. If the dowel is dry, it is time to give your citrus a thorough watering. Give your plant water until the water comes out of the drainage hole. Let the plant soak up the excess water for approximately thirty minutes, and if there is any left, use a turkey baster to draw the water out of the saucer.

FLOWERS: The flowers of citrus are white with yellow centers and have a lovely, sweet aroma. Use a paintbrush to brush the flowers to help pollinate them for fruit production. Lemons are self-pollinating, but as there are no insects or wind inside to help with that, a paintbrush can take their place.

PROPAGATION: Citrus can be grown from seed, but it may be years before any flowers or fruit are produced. If you take cuttings from a mature, blooming plant, you will have flowers much sooner. Place the cuttings in moist potting medium, propagating in the spring for best results.

TOXICITY: The essential oils in the leaves and fruit rinds are toxic to cats and dogs.

Clerodendrum thomsoniae

Bleeding-Heart Vine, Butterfly Flower, Glory Bower, Pagoda Flower

Clerodendrum is not a plant often seen in the home but is becoming more easily accessible as it is often sold as a hanging basket in the spring. This energetic twining vine, originating in tropical African and Asia, is beautiful when it is in full flower. It was introduced by William Cooper Thomson (1829–1878), a Nigerian missionary who named it after his late wife. It does grow vigorously but can be kept smaller by keeping it pruned. *Clerodendrum thomsoniae* blooms on new growth, so it is better to keep trimming the longer stems to promote new shoots and prolific flowering. In a hanging basket the flowers can be enjoyed from below, but it can also be trained around a trellis if you prefer it that way. You may find other *Clerodendrum* at the garden center that will interest you. You may find *Clerodendrum splendens* instead of *C. thomsoniae*. It resembles its relative except the white part of the flower is red, so the contrast is not there, yet it is still showy. There is also the *Rotheca myricoides* 'Ugandense' (formerly *Clerodendrum ugandense*) or blue glory bower. It's butterfly-like flowers are sky blue.

LIGHT: To successfully get this plant to bloom, it needs as much light as you can provide. A south or west window would be the best place, but if it still refuses to bloom in that light, it may need some supplemental light to flower. The shorter days of fall into winter will help trigger the flowering of *C. thomsoniae* in the spring, so try to keep it from supplemental light after dark.

WATERING: Plant this vine in a well-drained potting medium. Allow it to dry down before watering again, but don't allow it to dry out completely. If it is allowed to dry out, the result will be dropped leaves. It will do the same if the temperature gets much below 65°F (18°C), but it will leaf back out when it warms up. This plant loves extra nutrients and may become chlorotic if not fertilized regularly during the growing season.

FLOWERS: As the name implies, the flower resembles a heart with red "blood" flowing from the bottom of it. The flower petals or corolla are red, but the white part of the flower isn't a flower at all, but the calyx—the collective name of the four white sepals that protect the red flower bud. As the flower ages, the sepals will become a pinkish purple color, extending the season of color for this flowering plant.

PROPAGATION: This vine can be propagated from tip cuttings. Cut 4 to 6 inches (10 to 15 cm) off the end of a stem and either root it in water or plant it into a container of moist potting medium. To make a fuller plant faster, take a few cuttings to fill out the pot faster.

TOXICITY: *Clerodendrum thomsoniae* is not toxic to pets.

Clivia miniata

Natal Lily

The leaves of this plant are dark green and strappy and not all that exciting. Yet, when the vivid orange flowers arise out of the leaves in late winter, the foliage disappears into the background. The flowers are striking. The plants can become quite large, reaching heights of 2 feet (60 cm) high and 3 feet (1 m) in width. Each fan of leaves will send out flowers when they are mature.

This plant was named in honor of Lady Charlotte F. Clive, Duchess of Northumberland (1787–1866). She was a plant enthusiast, born into a plant-loving family and reputably the first person to get this South African native to bloom in her greenhouse in Great Britain.

LIGHT: Give *Clivia* plenty of bright light. In its natural habitat in the tropics of South Africa, it would receive dappled sunlight through the trees it grows underneath. In a home, place it in an east or west window or back 1 foot (30 cm) or so from a south window. If it develops brown spots on the leaves, it may be receiving too much sun. If the leaves become weak and floppy, it may not be receiving enough light.

WATERING: This plant has a hefty root system with thick roots. Give it a well-drained potting medium so the roots are supplied with plenty of oxygen. Add coarse orchid mix to a regular houseplant potting medium. This plant is drought tolerant because of its thick roots, but don't let it completely dry out. In the late fall and winter, it should be allowed to rest, and water should be minimal.

FLOWERS: The umbels of flowers are produced on tall stalks that arise out of the center of the strappy leaves. They are primarily vivid orange, trumpet-shaped flowers with yellow throats, but *Clivia* cultivars are also available in yellow and darker shades of orange. The key to getting the *Clivia* to bloom is to give it a dry, cool time of rest in the winter. If it is outside for the summer, wait to bring it in until the nights are falling into the 50°F to 60°F (10°C to 15°C) range. If you can place it in a cooler room or basement with minimal light and water, it should burst into bloom in the late winter.

When the flower stalk is visible down among the fan of leaves, wait until it is approximately 6 inches (15 cm) long before resuming regular watering, so it doesn't bloom down inside the leaves. If kept on the cooler side, the flowers are long-lasting. Don't be hasty to up pot up your *Clivia*. It is a plant that prefers to be snug in its pot—which helps stimulate flowering.

PROPAGATION: After the flowers fade, seeds are produced that will ripen into red berries. These can be used to start new plants, but seedlings may not flower for years when grown from seed, so dividing the plant preferably when it is root bound is the best way to propagate it. *Clivia* don't like to be disturbed, so wait until it needs to be up-potted before disturbing it.

TOXICITY: *Clivia* are toxic to cats and dogs.

Gardenia jasminoides

Gardenia, Cape Jasmine

Gardenia is a plant that brings to mind warm southern nights and large corsages of white flowers. It is a staple of the southern landscape, often used as a hedge. Its glossy, dark green foliage along with the creamy white flowers, which smell heavenly, make it a must-have in our outdoor summer garden and our inside garden during the winter. Make sure to place it where the heady scent of the flowers can be enjoyed. If given the right conditions, *Gardenia* will bloom all winter, filling a room with its lovely aroma.

This plant is named after Scottish physician and botanist Dr. Alexander Garden (1730–1791), who was a friend of Carl Linnaeus. *Gardenia* plants originate in tropical and subtropical parts of Africa, China, Australia, and Japan.

LIGHT: This plant will appreciate as much light as you can give it in the indoor garden. Though it has been said that they shouldn't be in full sun, it is fine in homes. Place it in a west or south window or supplement it with electric light if the light levels are too low in your home to allow it to bloom.

WATERING: *Gardenia* like to be evenly moist. They prefer an acidic soil, and a pH that is below 6.0 is best, so they are usually potted up in a high peat content medium. If the pH is too high, the leaves will yellow, and the plant will appear sickly. *Gardenia* grow better with high humidity so place them on a pebble tray under the plant or use a humidifier in the room.

FLOWERS: The creamy white flowers of *Gardenia* look like textured white roses, and they smell equally lovely, if not better. If you would like larger flowers, pinch the new growth on either side of the flower bud to send more energy to the single bud. The unfortunate thing is that when the flowers are spent, they turn a papery brown color that is unattractive and should be cut off at that point. You can cut the flowers for corsages or bouquets, being careful not to handle the flowers too much as it may cause brown spots on the petals if bruised. *Gardenia* flower best at temperatures below 75°F (24°C), doing best around 65°F (18°C). If they already have formed buds and the temperature falls too low or rises too high, the buds will drop. These plants have been touted as finicky plants, but with the right care, they aren't too fussy.

PROPAGATION: To propagate *Gardenia*, take tip cuttings in the spring. Clip off a cutting about 3 to 4 inches (8 to 10 cm) long, dipping it in rooting hormone, and plant it in small pots of moistened potting medium. Cover them with a glass cloche or a plastic bag to keep the humidity and heat up until they have rooted.

TOXICITY: *Gardenia* is toxic to pets.

Justicia brandegeeana (formerly *Beloperone guttata*)

Shrimp Plant, False Hop

When this plant sends out its dark peach-colored curved bracts, the reason for its common name becomes evident. Each individual bract is heart shaped and partially lays over the next bract, creating a layered look not unlike a hop plant inflorescence. The bracts at the end of the inflorescence are light green and darken to the peach with age. The true flowers consist of two small white petals peeking out from those lighter bracts. *Justicia* can become leggy with age, so keep it pruned to ensure it stays compact, allowing the branches to support the inflorescences.

The shrimp plant genus is named for Sir James Justice (1698–1763), a Scottish horticulturist, and the species honors Townshend Stith Brandegee (1843–1925), an American botanist who was interested in the plants of California and Mexico where the plant originates.

LIGHT: Though this plant grows in shady tropical areas, it will need a bright light but not direct sun in your home—an east or west window is preferable. If it refuses to send out any blooms, move it into more light. If the leaves are pale or develop brown spots, it may be in too much light, so move it farther away. The goal is to keep it compact and sending out blooms regularly. If you are lucky to have the variegated version of the shrimp plant, it will need brighter light to keep its variegation. Make sure to turn your plant regularly to give all sides of the plant light so the flowers appear on all sides.

WATERING: Keep the shrimp plant evenly moist but not saturated. Plant it in a well-drained potting mix so that the roots don't rot from being in too-moist soil. If it is too dry or too wet, it will drop leaves. Elevate the humidity in the winter when the heat is running to keep the spider mites from taking up residence on your plant as they love dry air and dry, stressed plants. If allowed to dry out, the thin leaves of this plant will wilt quickly and may not recover if kept dry for too long.

FLOWERS: The peach-colored bracts of the shrimp plant are really modified leaves. The bracts are part of the inflorescence that will attract pollinators that the plant needs. The small, white flower has purple spots on its lower petal that are like small airport runway lights guiding the pollinator to the nectar. In its native habitat, that would be a hummingbird, butterfly, or bees.

PROPAGATION: Propagate your shrimp plant by taking stem-tip cuttings. Pot them into a moist potting medium and cover them with plastic (or place them in a cloche) to keep the humidity elevated. The leaves are paper thin, and the cuttings will root faster if the humidity is kept up, keeping the leaves from wilting. At the same time, don't allow the condensation to get too high inside or it may rot the leaves.

TOXICITY: The toxicity of this plant to pets was not available.

Medinilla magnifica

Rose Grape, Chandelier Tree, Malaysian Orchid

When *Medinilla* is in full bloom with its pink inflorescences hanging from its branches, I can picture a beautiful chandelier in my mind. When the small flower petals have fallen and the developing seeds are left behind, a cluster of rose-colored grapes comes to mind. See how these common names are given to this plant? Although the botanical name is exclusively for one particular plant, in this case it still uses names that describe the habit or appearance of the plant in some way. In this case, the species *magnifica* does describe how magnificent this plant is in full flower.

Native to the Philippines, *Medinilla* was named in honor of José de Medinilla y Pineda, governor of Marianne Islands (now Mauritius), off the coast of Africa. In this tropical setting, the *Medinilla* grows as an epiphyte in the crooks of trees. Imagine coming upon this plant with its flowers dripping from huge trees above your head! Consequently, *Medinilla* have root systems that need plenty of air, so use a well-drained potting mix with plenty of orchid bark added. Plant it in a shallow clay pot so the roots are better able to get oxygen and excess moisture will evaporate quickly. The large leaves should be kept free of dust and the humidity elevated to keep spider mites from making your plant their next meal.

LIGHT: Since this plant is naturally an epiphyte in the Philippines, it is receiving dappled light from its perch high in the trees. In your home, an east or west window would be sufficient light for this plant. A south exposure with direct sunlight might burn the leaves. If it doesn't send out flowers in the spring, move it to a higher light location.

WATERING: This plant needs thorough watering but does not want to stand in water or be overly wet. This could cause root rot. The key is raising the humidity, and that can be done with a large pebble tray filled with water sitting under the plant saucer. When you water *Medinilla*, take it to the shower to give it added moisture and rinse dust off its large leaves. Fertilize this plant regularly to keep it healthy.

FLOWERS: The true flowers are pale pink panicles dangling beneath large pink bracts, the showiest part of the inflorescence. After the petals fall, the developing seed pods are also a showy pink, making for a long season of color.

PROPAGATION: Because this plant can become woody with age, young stem cuttings would be the best way to propagate it. Use a rooting hormone, and as is often done with large-leaved plants, cut up to half of the leaf off so that the cutting isn't losing as much moisture. The seeds can also be used to start new plants, but it takes two to three years to get a plant to the flowering stage.

TOXICITY: This plant is not toxic to cats or dogs, but as with any plant, do not allow them to nibble on the leaves.

Spathiphyllum

Peace Lily, White Sails

This plant is best known as a florist plant and is often part of large basket arrangements or as a gifted memorial plant. Though this may sound depressing, it is wonderful to receive a plant that reminds you of a special person. The popularity of *Spathiphyllum* comes from their white inflorescences, which remind us of purity and peace. The genus name refers to the Greek words *spathe* meaning "blade" and *phyllon* meaning "leaflike." This plant has this name because the part that is assumed to be the flower is a spathe or modified leaf that protects the true flower or spadix, which is the knobby spike nestled inside the spathe. It is also white and often the pollen that falls from the spadix is excessive, sprinkling the leaves below with what looks like white powder. This phenomenon may alarm you as you may think you have some sort of insect invasion. Simply brush or wipe the excess pollen off with a damp cloth.

This plant comes in sizes from 1 foot (30 cm) in height to upward of 5 or 6 feet (1.5 to 1.8 m) with enormous leaves and inflorescences. These larger forms aren't often found to purchase for obvious reasons. Keeping this plant healthy and blooming isn't difficult in our homes and is perfect for that lower light area.

LIGHT: The light needed for this plant isn't excessive, and in fact, it doesn't want to live in a full sun area. If it is in direct light, the thin, dark leaves will suffer from sunburn. Instead, a bright light from an east or west window is better.

WATERING: Keep this plant moist at all times. Some people use the wilting of this plant to indicate its water needs, yet this is not the best practice. *Spathiphyllum* does come back from wilting, but with consequences, those being yellowing leaves, brown leaf tips, and loss of flowers. It is better to keep it moist and if possible, with elevated humidity. These tropical plants love warmth, humidity, and moist soil. Place it on a pebble tray or use a humidifier in the vicinity of the plant.

FLOWERS: As mentioned above, the actual flowers of the peace lily are surrounded by a spathe that is a modified leaf. This large, white flag is a sign to passing pollinators that a flower is present and needs pollination. The spadix is a large collection of flowers all packed together on one spike. These produce copious amounts of pollen, which can be seen falling when the flower is touched. As the flower ages, it turns green and then brown. Cut off the spent flower stem as close to the base as possible.

PROPAGATION: The best way to propagate the peace lily is by division. It grows in individual plantlets with multiple plants in each pot. As the plant gets older, more of these individual plants will emerge to fill the pot. Take the plant out of the pot and carefully pull the plantlets apart, or if that isn't possible, a knife may be needed. Separate the plants and pot each one into an individual container.

TOXICITY: *Spathiphyllum* is toxic to pets.

Resources

Plants

Andy's Orchids
www.Andysorchids.com
Orchids and other indoor plants.

Baker's Acres Greenhouse
www.bakersacresgreenhouse.com
Houseplants, including many rare finds.

Bloomscape
www.bloomscape.com
Mail-order houseplants and tools.

Gabriella Plants
www.gabriellaplants.com
Large selection of houseplants, including unique and rare plants.

Glasshouse Works
www.glasshouseworks.com
Never-ending selection of houseplants.

Groovy Plants Ranch
www.groovyplantsranch.com
Many unusual houseplants as well as the humble standbys.

In Succulent Love
www.insucculentlove.com
Mail-order nursery for succulent plants.

Josh's Frogs
www.joshsfrogs.com
Plants as well as frogs and other animals.

Kartuz Greenhouses
www.kartuz.com
Large selection of plants, including flowering plants, gesneriads, and begonias.

Little Prince of Oregon
www.littleprinceplants.com
Great houseplants as well as outdoor plants.

Logee's Tropical Plants
www.logees.com
Over 100-year-old greenhouse offering houseplants with an amazing begonia selection.

Lyndon Lyon Greenhouses
www.lyndonlyon.com
Offering African violets and other gesneriads.

The Sill
www.thesill.com
Mail-order indoor plants and planters.

Steve's Leaves
www.stevesleaves.com
Mail-order company with exotic and rare houseplants.

The Violet Barn
www.violetbarn.com
African violets, gesneriads, and small terrarium plants.

Lights and Containers

Gardener's Supply Company
www.gardeners.com
Lights, stands, and supplies for gardeners.

Puddingstone
www.pstone.space
Handmade terra cotta pots as well as kokedama (moss ball) houseplants.

Soltech Solutions
www.soltechsolutions.com
Lights for growing your houseplants if you don't have good ambient light in your home.

Acknowledgments

I grew up in a rural area of mid-Michigan and fortunately, my paternal grandparents lived only a half mile (1 km) away. I rode my bike there often and spent a lot of time with my grandma. She was the best housekeeper (I did not inherit that) and cook and had a green thumb. I did inherit that. As a hardworking farmer's wife, she had a huge garden full of veggies that she canned and froze, preserving them for the winter. But the garden was also overflowing with beautiful flowers. Her dining room table often sported a gorgeous flower arrangement, and her windows were filled with blooming African violets. I loved those violets and watching Grandma take care of them. She also had amaryllis and an *Aechmea* like the one on the front cover, my first introduction to bromeliads. So, my first thank you goes to my grandma, Rose Elnore (Lyon) Eldred, for instilling in me a love for flowering houseplants. Also, to my mom, Christine Ann (Baldwin) Eldred, for starting me on the path to loving all houseplants with her amazing fern, which I now care for. A part of this fern has now been passed along to the third generation of plant-loving Eldred/Steinkopf women.

I have many other people to thank. Thanks go to my best friend Jeanine Merritt for her support and for being my biggest cheerleader. Also, to my daughter Hayley Steinkopf Bonafede for being my model for many of the pictures and to her and my son-in-law, Marcos Bonafede, for opening their home and cottage to stage pictures. To my daughter Chelsea Steinkopf for her priceless photo expertise. To Erin Hannum for her camera instruction. To Rachel Nisch of Graye's Greenhouse for allowing me to take pictures in her magical greenhouse. To Tim Travis of Goldner Walsh Garden & Home for also allowing me to take pictures. To Lynn and Doug Allen for letting me take pictures at their home and to Lynn for her extensive African violet knowledge generously shared. To my brother Keith Eldred, Michigan inspector for MDARD, and Chris Baker, of Baker's Acres, who gave me some great plant people contacts, including Rick Schoellhorn who led me to another great contact. It's good to know people who know people! Thank you to Dr. Erik Runkle, professor at Michigan State

University, for his time and knowledge. To good friend Susan Martin of Gardener Sue's News, who was not only helpful with editing but also moral support. To friend Nancy Szerlag for sharing her plant and soil expertise. To friend Laura Mittlestat King, a middle school English teacher who was so helpful with grammar questions.

A big thank you to Paul Wingert, President of the Southeast Michigan Bromeliad Society, not only for sharing his vast knowledge of bromeliads but also for allowing me to use his plants for pictures and for editing the bromeliad information. To Mel Grice, President of the Gesneriad Society of America, for editing the gesneriad plant profiles. To Dave Hinch, Tom Dunkowski, Tony Wamsley, Brian Riley, and Carol Koss from the Michigan Cactus and Succulent Society for their plant expertise.

To my editor at Cool Springs Press, Jessica Walliser, who is a dream to work with. Also, to Art Director Heather Godin, Project Manager Elizabeth Weeks, and copy editor Anne Marie Van Nest for their hard work making my book beautiful. Thank you!

To my friends and family who haven't seen me much this year, to my heavenly father who has made all of this possible, and finally to my fabulous best friend and husband, John, for all his patience and support as I immersed myself in this book-writing venture. Someday, the house will be clean and there will be less plants. Love you!

About the Author

LISA ELDRED STEINKOPF, The Houseplant Guru, has been obsessed with houseplants as long as she can remember and cares for 100s in her home. Her love for plants started during childhood, growing up immersed in nature in the farmland of mid-Michigan with her equally plant-obsessed brother, Keith.

She attended Ferris State University in Big Rapids, Michigan, to study horticulture. While there she met John, now her husband of thirty-seven years, who is an owner of his family's ninety-year-old, fourth-generation garden center. Plants are obviously a large part of her and her family's lives.

Lisa features all things houseplants on her website, www.thehouseplantguru.com. She has written three books, *Houseplants: The Complete Guide to Choosing, Growing, and Caring for Indoor Plants*; *Grow In The Dark: How to Choose and Care for Low-Light Houseplants*, and *Houseplant Party: Fun Projects and Growing Tips for Epic Indoor Plants.* She has written for many magazines, including *Horticulture, Real Simple, Fine Gardening, Better Homes and Gardens, First for Women*, and HGTVgardens.com. She has also written the houseplant section for Allan Armitage's Greatest Perennials and Annuals app. Because of her love of outdoor gardening as well, she writes a column for *Michigan Gardener* magazine. She has been interviewed on radio, television, in print, and on numerous podcasts such as Bloom & Grow Radio, On The Ledge, and Epic Gardening.

She is a member of many plant societies—including the Michigan Cactus and Succulent Society, of which she is a board member, the Town and Country African Violet Society, the Southeast Michigan Bromeliad Society, Garden Communicators International, and the Hardy Plant Society. Plant societies are a great place to meet like-minded plant geeks—join one!

She loves to visit conservatories and gardens during her travels and volunteers at the Belle Isle Anna Scripps Whitcomb Conservatory in Detroit.

Lisa feels everyone should have a plant or two in their home and office because there is a plant for every situation, especially if grown under electric lights. She lectures extensively around the country about her love of houseplants and demonstrates how everyone can have some in their home and be able to care for them. A green thumb is something everyone can have because we all need a little green in our lives!

Index

Q

R

S

T

More books from Lisa Eldred Steinkopf

Houseplants
The Complete Guide to Choosing, Growing, and Caring for Indoor Plants
978-1-59186-690-9

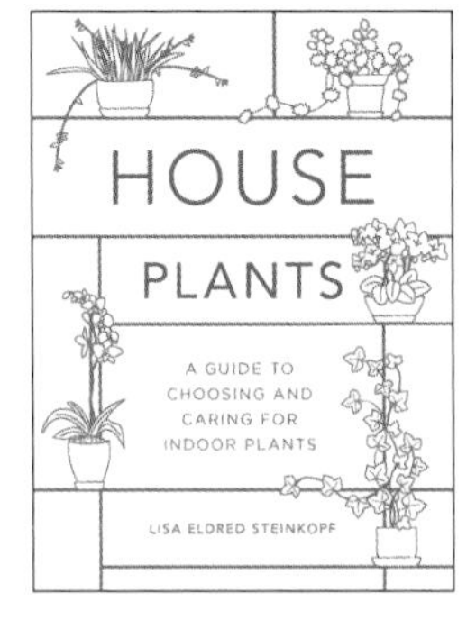

Houseplants (mini)
A Guide to Choosing and Caring for Indoor Plants
978-0-7603-6592-2

Houseplant Party
Fun Projects & Growing Tips for Epic Indoor Plants
978-1-63159-883-8

Grow in the Dark
How to Choose and Care for Low-Light Houseplants
978-0-7603-6451-2